I dedicate this book to my parents, grandparents, relatives, friends, and teachers.

K.R.

The Geography Bee Ultimate Preparation Guide

ISBN-10: 1511714824

ISBN-13: 978-1511714822

Printed in the United States

© Copyright 2015 Keshav Ramesh

Font Set in Maiandra GD, Arabic Typesetting, Candara, Calibri

Summary: This is a book designed to prepare students competing in the classroom, school, state, and national levels of the National Geographic Bee.

Design and Text by Keshav Ramesh

Cover Illustration by Keshav Ramesh

All rights reserved.

THE GEOGRAPHY BEE ULTIMATE PREPARATION GUIDE

by Keshav Ramesh

The Geography Bee Ultimate Preparation Guide

Table of Contents

Tips, Tricks, and How to Prepare for the Geography Bee 6

United States 27

North America 65

South America 81

Asia 96

Europe 148

Africa 180

Australia and Oceania 200

Antarctica 206

Physical Geography 212

Cultural Geography 227

Economic Geography 241

Historical Geography 256

Political Geography 266

Animal/Environmental Geography 274

Current Events 280

National Parks/Forests and Sites 283

Simulation Bee 290

The Geography Bee Ultimate Preparation Guide

USA Geography Olympiad/iGeo Resources 315

Geo Statistics 318

About the Author..337

Acknowledgements 338

Links 340

Bibliography..341

Notes 344

The Geography Bee Ultimate Preparation Guide

Tips, Tricks, and How to Prepare for the Geography Bee

What materials should I have when studying geography?

When studying for the geography bee, I would recommend keeping an atlas and a detailed political map near you. These two are the most vital sources you need to help you achieve more in the competition.

From looking at a detailed political map, you can find cities and countries from around the world. Territories, islands, and dependencies will be included. You can find oceans, seas, lakes, straits, gulfs, bays, and sometimes even rivers. Archipelagos are scattered across the world, like Indonesia and Japan. A political map will show you the countries in different colors, and they will be labeled. Some political maps will also show you the continents, and the names of countries will be printed much bigger than the cities. A physical map will also help as well, to identify the landforms (which will be labeled) as well as biomes. Tectonic plate names and ocean ridges are sometimes shown, but not all the time. Different Earth colors will be expressed on the map, as well as ridges to show mountains, or a plain, yellowish color spreading across a certain area to signify a desert.

Atlases are great for researching thousands of facts just by looking at maps and diagrams, and reading about them. Countries are always featured in atlases, and you'll always find a few chapters about the physical geography of the world. I would recommend buying atlases sold by National Geographic, as they, in my opinion, provide the best information.

You should get an almanac, like the National Geographic Kids Almanac 2016 or another good one recently published with a lot more facts on geography that you can absorb.

You should also get a book about the physical geography of the world, as this is vital to the physical geography section of the National Geographic Bee.

How hard are the questions in this book?

The difficulty of these questions vary from the high reaches of the School Geography Bees to the preliminary rounds of the National Geographic Bee. You should use these questions to prepare to win your School Geography Bee, participate/win your State Geography Bee, and participate in the National Geographic Bee.

The questions increase in difficulty as you finish more and more, before matching questions found at the final levels of the state bee and the preliminary rounds of the national bee. On a scale of 1 to 10, 1 being the questions in the classroom bee and 10 being the questions in the finals of the national bee, this guide focuses between 6 and 10.

The Geography Bee Ultimate Preparation Guide

How many questions are in this book and how can I study using this book?

The Geography Bee Ultimate Preparation Guide has **over 1,930 questions to help you in then classroom, school, state, and national bees,** and even international geography competitions (Check out the USA Geography Olym-piad/iGeo chapter!).

You'll notice there are different chapters based on different aspects of geography. There's a chapter on the geography of the United States. World Geography is split into 7 chapters: North America, South America, Asia, Europe, Africa, Australia and Oceania, and Antarctica.

Human Geography is split into four chapters: Cultural Geography, Economic Geography, Historical Geography, and Political Geography. Physical Geography is a very large chapter by itself, comprised of Earth Science terms and information. Animal Geography focuses on animals and their habitats around the world, and the questions in this chapter will have geographical clues.

The chapter on Current Events will have questions as well as study tips, so use both to your advantage! There is also a section called National Parks/Forests and Sites, which focuses on National Parks, National Forests, Wildlife Reserves, National Monuments, and World Heritage Sites around the globe.

If you're preparing for the NGB but not the USAGO or iGeo, you should still check out the links posted in the USAGO chapter. There are hundreds of questions for you to prepare with!

What is the daily amount of time I should study?

I would recommend 30 minutes to 1 hour a day, or 1 hour in total every two days if you are preparing to win the school competition. For the state qualification test, 1 hour a day is necessary to almost guarantee you'll make it to the state level – the score was determined out of 60 points. For the state level bee, I would recommend studying 1 ½ to 2 hours a day.

If you're at the level of the National Bee Preliminaries or higher, 2 to 3 ½ hours is probably somewhere around your goal. On weekends, try to add an extra hour or two.

This competition, like the Scripps National Spelling Bee, National History Bee, MathCounts, and AMC 8 is challenging. Hours of dedication to geography is vital to your success in the state and national bees.

Where do the questions in this book come from?

Most of the questions in this book have been created by me and also from my mom when she was formulating questions from atlases to quiz me. A variety of questions come from National Geographic's GeoBee Challenge, a free game on the National Geographic website as well as one you can purchase on an app store.

There are a few ones which come from past bees, to give you a sense of the type of questions. I won't be telling you what questions those are, however. It's up to you to study all of them and treat them all like actual Bee questions, a great study skill for a competition.

What are the rules and FAQ for the National Geographic Bee?

Here is a link to the rules for the National Geographic Bee as of 2015. Ask your parents to read it with you:

http://www.nationalgeographic.com/geobee/rules/

Here is a link to the National Geographic Bee's FAQ:

http://www.nationalgeographic.com/geobee/frequently-asked-questions/

What should I study?

For your classroom competition, you need to know the basic facts about countries around the world and the United States: Locations, capitals, major rivers, bordering countries, major island countries, territories, and major water/land features.

This you probably know much about, if you have studied geography a little before. If you're aiming at the school competition, it is vital to know major rivers, mountains, and other physical landforms around the world. Capitals of countries and states must be known, as well as major cities throughout the world and the United States.

You will need to know a lot of Cultural Geography, and will need to research religion, language, indigenous cultures, and nationwide/global cultures. Economic geography is important too. The economy is an important part of human geography, and various questions will be asked throughout the competition about production, exports/imports, livestock, gross domestic product, etc.

What do I need to know about each country?

Here's a list of what you need to know about each country to help you prepare for the National Geographic Bee:

Basics:
- Location (Continent)
- Location (Region, e.g. South Asia)

- Capital
- Major Cities (At least 5-15, 15+ for countries with populations over 70,000,000)
- Population (Approximate)
- Population Density (Approximate)
- Area (Approximate)
- Official Country Name

Physical:
- Highest Point
- Lowest Point
- Mountain Ranges
- Major Peaks
- Rivers
- Lakes
- Bordering Seas
- Bordering Oceans
- Gulfs
- Bays
- Sounds
- Deltas

- Plateaus
- Grasslands
- Prairies
- Plains
- Basins
- Swamps
- Marshlands
- Depressions
- Deserts
- Valleys
- Passes
- Straits
- Waterfalls
- Dams
- Canyons
- Capes
- Spits
- Peninsulas
- Islands
- Canals
- Physical Regions

- Buttes
- Mesas
- Glaciers
- Inlets
- Archipelagos
- Isthmuses
- Volcanoes
- Major Reservoirs
- Lagoons
- Reefs
- Divides (Continental)
- Earth Physical Structure
- Atmosphere
- Layers of the Earth

Political

- Country Independence
- Bordering Countries
- Administrative Divisions (States, Provinces, Federal Districts, Counties, Parishes (State & National Levels))

- Territories
- Dependencies
- Occupied Atolls
- Current Leader(s) (President, Prime Minister, King/Queen, Chairman, Prince, etc.)
- Government Structure/Important and Influential Laws
- Type of Rule (Republic, Democracy, Monarchy, etc.)
- Kingdoms and Empires
- Disputed Countries
- Disputed Regions
- Politically Established Regions (Northeast Africa, Central America, etc.)
- Country and Territory Boundaries
- National Organizations
- Global Organizations
- Human Rights Organizations
- Organization Headquarters
- Defense and War

Cultural

- Religions
- Languages
- Currency
- Festivals
- Holidays
- Traditions
- Food
- Customs
- Native/Indigenous Tribes
- Art and Music
- Mythology
- Statues
- Architecture
- Cultural Items/Objects
- Cultural/Famous Symbols
- Language Families
- Language Groups

Environmental

- Climate
- Biomes

- Animals
- Plants
- Habitats
- Global Warming/Climate Change
- World Vegetation
- Environmental Hot Spots
- Natural Disasters

Economic

- Trade
- Production
- Major Exports
- Major Imports
- Gross Domestic Product - GDP
- Agricultural Products
- Natural Resources
- Resource Use
- Computer/Technology Production
- Computer/Technology Exports
- Computer/Technology Imports
- Livestock/Domestication

Historical:

- Kingdoms
- Empires
- Wars
- Civilization
- Ancient Culture
- Countries that No Longer Exist
- Middle Ages
- Dark Ages
- Historical Colonies and Territories
- Past Country Independence
- Historical Leaders
- Historical Type of Rule

Landmarks

- National Parks
- National Forests
- State Parks
- State Forests
- UNESCO World Heritage Sites
- Castles and Ruins

- Famous Museums
- Architecture
- Famous Zoos
- Animal Sanctuaries
- Wildlife Reserves
- Protected Areas
- Seven Wonders of the World
- Seven Wonders of the Ancient World
- World Landmarks
- United States Landmarks
- Native American Reservations
- Space Centers and Observatories
- National Laboratories
- National Wildlife Refuges

Current:

- Global Issues
- Nationwide Issues
- Environmental Issues
- Current Disputes
- Current Disasters

- New Discoveries
- Archaeological Discoveries
- Global News
- Nationwide News
- Climate Change and Global Warming
- Treaties, Pacts, and Agreements
- Diplomatic Relations
- Major Worldwide Sports (2014 FIFA World Cup, etc.)

Will I need to study history for this competition too?

The National Geographic Bee is nothing like history competitions you might hear of or participate in, except that they are both competitions with questions that give you clues. However, the National Geographic Bee will ask questions pertaining to geographical history such as:

- Kingdoms
- Empires
- Wars
- Civilization

- Ancient Culture
- Countries that No Longer Exist
- Middle Ages
- Dark Ages
- Historical Colonies and Territories
- Past Country Independence
- Historical Leaders
- Historical Type of Rule

Just in case, study these. It will not come up often, but these questions can sometimes fool you.

What are some good websites I can use to prepare?

The website I'd recommend to use for geography bee preparation is www.nationalgeographic.com/geobee. It has great tools to help you study, and also a page where you can play a game called the GeoBee Challenge, where National Geographic gives you 10 new questions every day to prepare with. This is a list of all of the websites you should use:

www.nationalgeographic.com/geobee

www.geobeeworld.blogspot.com

www.prepgeobee.blogspot.com

www.kids.nationalgeographic.com

www.geography.about.com

www.geobeeprep2011.blogspot.com

www.world-geography-games.com/

www.lizardpoint.com/geography/index.php

www.ducksters.com/geography/

www.socialstudiesforkids.com/

www.worldatlas.com

www.cia.gov/library/publications/the-world-factbook/

www.mapsofworld.com

What books do you recommend I read to prepare and/or gain knowledge?

There are many books about geography that will enhance your skills besides this one. You should get National Geographic Kids World and United States Atlases, National Geographic Almanacs, Physical Geography and/or Earth Science books, and History books.

Does this book cover only the National Geographic Bee?

This is important!

This book not only covers the National Geographic Bee, but can be used for preparation in the National Geographic World Championship, an international geography competition where the top three geography students in each country compete to claim the title.

The Geography Bee Ultimate Preparation Guide is an **excellent resource for the North South Foundation's Junior and Senior Geography Bees.** This book is guaranteed to help you advance greatly in the NSF Geography Bees.

This book is also good for preparation for the USA/International Geography Olympiad. There is even a chapter dedicated to past USA/International Geography Olympiad Regional and National Questions.

Any questions or concerns?

The Geography Bee Ultimate Preparation Guide

Are there any errors needed to be fixed? Any questions you'd like me to add? Still need help with geography? Contact me, Keshav Ramesh, at keshav.ramesh@gmail.com or go to

www.prepgeobee.blogspot.com and comment in the Questions and Concerns post.

Visit www.geobeeworld.blogspot.com to get more preparation for the bee!

Visit the site for more questions, tips, and hints on how to win the school, state, and/or national levels of the National Geographic Bee!

The Geography Bee Ultimate Preparation Guide

United States

1. The Lewis and Clark Expedition explored which region from 1804 to 1806 – Louisiana Purchase or Southwest?
   Louisiana Purchase

2. Lake Sakakawea is a 200-mile-long (320-kilometer-long) reservoir along which U.S. river – Missouri River or Colorado River?
   Missouri River

3. What city, founded by William Penn, was the site of the First and Second Continental Congresses – Baltimore or Philadelphia?
   Philadelphia

4. Which one of the contiguous United States has the northernmost point – Minnesota or Wisconsin?
   Minnesota

5. Which city is more likely to have desert vegetation – Phoenix or Denver?
   Phoenix

6. Which city is located on the western tip of Lake Superior and is an important port for shipping grain and iron ore – Duluth or Chicago?

Duluth

7. What Alabama city was the subject of a 1955 and 1956 bus boycott organized by Martin Luther King, Jr., to protest bus segregation – Montgomery or Birmingham?
Montgomery

8. Which state has more coastal marshland – Louisiana or Kentucky?
Louisiana

9. Which city is in western Texas and lies directly across the border from the Mexican city of Ciudad Juarez – Amarillo or El Paso?
El Paso

10. Badlands National Park and Mount Rushmore are west of the Missouri River in which state – Minnesota or South Dakota?
South Dakota

11. Which northeast state borders Canada – New Hampshire or Massachusetts?
New Hampshire

12. Cape Cod, known for its beautiful shoreline, is located in which state – New Jersey or Massachusetts?
Massachusetts

13. When it's 1 p.m. in Hawaii, what part of the day is it in the state of New York – midnight or evening?

Evening

14. Which state produces more wood pulp – Georgia or Florida?
    Georgia

15. Land bordering the mouths of the James, York, and Rappahannock Rivers is part of the Tidewater region of which state – Virginia or Ohio?
    Virginia

16. Poi is a food commonly associated with people native to which state – Hawaii or Alaska?
    Hawaii

17. The melaleuca tree is believed to be drying up parts of a large swamp, making it uninhabitable to native plants and animals. This process is occurring in what swampy region – Great Basin, Everglades, or Pine Barrens?
    Everglades

18. Tornadoes occur most frequently in which region of the United States – Northwest, Southwest, or Midwest?
    Midwest

19. Which state has more national parks – California, Delaware, or Iowa?
    California

20. The Constitution State is the state nickname for what state – Connecticut or Maryland?
    Connecticut

21. The easternmost point in the contiguous United States is located in what state – Massachusetts or Maine?
Maine

22. El Paso is a city that lies close to Mexico in what state famous for its oil production – Texas or New Mexico?
Texas

23. Mount St. Helens is a volcano in what state – Oregon or Washington?
Washington

24. Miami, Orlando, and Tallahassee are all cities in what state home to the Everglades and Walt Disney World – Louisiana or Florida?
Florida

25. Chicago borders what Great Lake – Lake Michigan or Lake Huron?
Lake Michigan

26. Florida, Texas, Louisiana, Alabama, and what other southeastern state borders the Gulf of Mexico – Mississippi or Georgia?
Mississippi

27. The Great Miami River and Little Miami River are two rivers in what U.S. State in the Eastern Time Zone – Florida or Ohio?
Ohio

28. The Golden Gate Bridge is located in what state that is home to the cities of Cupertino and Oakland – New Jersey or California?
California

29. Cape Canaveral is located in what state – Florida or Massachusetts?
Florida

30. The port of Mobile is located in what U.S. State – Mississippi or Alabama?
Alabama

31. In which state is more land used for agriculture – Connecticut or Illinois?
Illinois

32. Which state borders California on the north – Oregon or Utah?
Oregon

33. In July 2002, the U.S. Congress voted to authorize Yucca Mountain as a permanent repository for 77,000 tons of nuclear waste. Yucca Mountain is located in which state – Kansas or Nevada?
Nevada

34. The deep roots of the tamarisk, or salt cedar tree, can deprive native plants of water. Although concentrated in

the arid southwest, this tree has spread as far as what state located north of Wyoming – Idaho or Montana?
Montana

35. Newport, a prominent seaport in colonial times, is now a popular tourist area located in which state – Rhode Island or Delaware?
Rhode Island

36. Hydrilla, introduced to Florida as an aquarium plant, now clogs many waterways in the United States. This plant, which is spread easily by boats, has been found near Washington, D.C., in which river – Hudson or Potomac River?
Potomac River

37. Brownsville, an important border city on the Rio Grande, is located in which state – New Mexico or Texas?
Texas

38. Which state does not have a coastline – New Mexico or New Hampshire?
New Mexico

39. Which bay has a larger total area – Prudhoe Bay or Chesapeake Bay?
Chesapeake Bay

40. The Brooks Range is located in what state – Maine or Alaska?
Alaska

The Geography Bee Ultimate Preparation Guide

41. Mount Whitney is the highest point in the contiguous United States, located in what state – California or Nevada?
California

42. The Colorado Plateau, located in the southwestern United States, is between the Rocky Mountains and what else – the Great Basin or Ouachita Mountains?
The Great Basin

43. The Chilkoot Pass crosses from Alaska into what country – Canada or Mexico?
Canada

44. The Everglades, a large swamp, can be found in the southern part of what state where Walt Disney World can be found?
Florida

45. The San Joaquin Valley is located between two mountain ranges. These mountain ranges are the Sierra Nevada and what other range – Coast Range or Cascade Range?
Coast Range

46. The Chihuahuan Desert, found in Texas and Mexico, extends into what southwestern U.S. state – Arizona or New Mexico?
New Mexico

47. The United States borders an amazing three oceans. These oceans are the Atlantic, Pacific, and what other ocean

including the Beaufort and Chukchi Seas – Arctic or Southern?
Arctic Ocean

48. Puget Sound feeds into the Pacific Ocean. This sound is located in what state whose capital is Olympia and is home to the Seattle Seahawks football team?
Washington

49. The Arkansas River, a tributary of the Mississippi River, has its source in what major mountain range – Rocky Mountains or Appalachian Mountains?
Rocky Mountains

50. The Great Salt Lake is located in what state home to large populations of Mormons – Utah or Wyoming?
Utah

51. Lake Okeechobee, on the edge of the Everglades, is located in what state with the cities of Orlando and Miami – Georgia or Florida?
Florida

52. Chesapeake Bay is an estuary of what river – Susquehanna River or Allegheny River?
Susquehanna River

53. Lake Pontchartrain is located on what plain – Great Plains or Gulf Coastal Plain?
Gulf Coastal Plain

The Geography Bee Ultimate Preparation Guide

54. Yosemite Falls, located in the Sierra Nevada, is in what region of the United States – southern or western?
Western

55. Cape Cod and Cape Canaveral are located on what coast of the United States – west coast or east coast?
East coast

56. Howland Island, Jarvis Island, and Wake Island are all territories of the United States located in what ocean – Atlantic Ocean or Pacific Ocean?
Pacific

57. People from Mexico and what other country constitute the highest number of U.S. immigrants every year – India or China?
India

58. Permafrost in the United States is found in what state – Montana or Alaska?
Alaska

59. Molokai is one of the main islands of what state that was the last to join the U.S. – Alaska or Hawaii?
Hawaii

60. Houston is a major port on what gulf – Gulf of Mexico or Gulf of Alaska?
Gulf of Mexico

The Geography Bee Ultimate Preparation Guide

61. The District of Columbia (Washington, D.C.) is located between Maryland and what other state – Delaware or Virginia?
Virginia

62. Cape Hatteras can be found off the coast of what state home to Duke University – North Carolina or South Carolina?
North Carolina

63. Long Island Sound is north of Long Island and south of what state home to the historical figures of Oliver Ellsworth and Noah Webster – Connecticut or Rhode Island?
Connecticut

64. Glen Canyon is a hydroelectric dam on what river – Monongahela River or Colorado River?
Colorado River

65. The Gulf of Maine feeds into what ocean – Pacific or Atlantic?
Atlantic

66. Lake Michigan is the only Great Lake entirely in the United States. This lake also borders what state where the cities of Lansing and Detroit can be found? – Michigan or Illinois?
Michigan

67. Which Great Lake is the smallest in size – Ontario or Erie?
Erie

68. The Saint Lawrence River forms part of the border between the United States and what other country – Mexico or Canada?
Canada

69. The Seward Peninsula is located in what state whose major cities include Anchorage and Fairbanks – Alaska or Washington?
Alaska

70. Phoenix, a major U.S. city and the capital of Arizona, is on the edge of what southwestern U.S. desert – Sonoran Desert or Chihuahuan Desert?
Sonoran Desert

71. Chicago is the chief port on what lake – Lake Michigan or Lake Superior?
Lake Michigan

72. Cape Flattery is bordered by the Pacific Ocean and what strait – Strait of Juan de Fuca or Bering Strait?
Strait of Juan de Fuca

73. The Aleutian Islands are a group of islands belonging to Alaska that extend across what meridian – 180 degree or 170 degree?
180 degree

74. Hurricane Katrina is famous for wrecking what major city – New Orleans or Mobile?
New Orleans

75. Amarillo is a city in what state famous for its oil production – Oklahoma or Texas?
Texas

76. The Straits of Florida connect the Atlantic Ocean to what gulf – Gulf of Alaska or Gulf of Mexico?
Gulf of Mexico

77. The Colorado River, emptying out into the Gulf of California, created what major canyon in Arizona – Grand Canyon or Palo Duro Canyon?
Grand Canyon

78. Hoover Dam can be found on the border between Arizona and what other state known as the Silver State – Nevada or New Mexico?
Nevada

79. The longest river in New England is what river – Connecticut River or Hudson River?
Connecticut River

80. Lake Itasca can be found in what state – Mississippi or Minnesota?
Minnesota

81. Hartford, known as the insurance capital of the United States, is the capital of what state where the first hamburger was served in U.S. history – Connecticut or Rhode Island?

Connecticut

82. Wisconsin borders what state to the east – Michigan or Illinois?
    Michigan

83. Louisville and Knoxville are cities in what state that straddles the world's longest Cave System, Mammoth Cave National Park – Kansas or Kentucky?
    Kentucky

84. Lake Champlain borders New York and what other U.S. state – Vermont or New Hamsphire?
    Vermont

85. Redwood National Park can be found in the northern part of what state – California or Oregon?
    California

86. Marquette and Flint are cities in what state bordering four lakes – Wisconsin or Michigan?
    Michigan

87. International Falls can be found in what state with the cities of Duluth and Rochester – Minnesota or New York?
    Minnesota

88. Cape Mendocino and Point Conception can be found in what state famous for the Sierra Nevada, Mojave Desert, and Mt. Whitney – Nevada or California?
    California

89. Which bay is connected to the Atlantic Ocean – Monterey Bay or Chesapeake Bay?
Chesapeake Bay

90. Which state has a longer border with Mexico – Texas or California?
Texas

91. Which city has a greater earthquake risk – Los Angeles or New York City?
Los Angeles

92. New York is the only state that borders which lake – Lake Erie or Lake Ontario?
Lake Ontario

93. Which state does not experience frequent earthquakes – California or Massachusetts?
Massachusetts

94. Cape Cod, Cape Hatteras, and Cape May are along which coast of the United States – West or East?
East

95. Which one of the 50 states shares the longest border with Mexico – New Mexico or Texas?
Texas

96. Which city is in southern Nevada and saw its population grow by more than 80 percent in the 1990s – Las Vegas or Renio?
Las Vegas

97. Water from the Snake River makes it possible for which mountainous state to produce more than one-quarter of the country's potato crop – Idaho or Wyoming?
Idaho

98. Albuquerque, New Mexico, is to the Rio Grande as Bismarck, North Dakota, is to WHAT – Snake River or Missouri River?
Missouri River

99. Which state capital is farther north – Augusta, Boston, or Hartford?
Augusta

100. Which city is located at a higher elevation – Denver or New Orleans?
Denver

101. Which state borders more states – Kentucky, Alaska, or Rhode Island?
Kentucky

102. The Pocono Mountains in Pennsylvania link with what mountain group to the northeast – Catskill Mountains or Adirondack Mountains?
Catskill Mountains

103. Which U.S. state is located closer to the Tropic of Cancer – Louisiana or Maine?
Louisiana

104. Which state has more people per square mile – Indiana or New Mexico?
Indiana

105. Redwoods, the world's tallest trees, are found in which state – California or Alaska?
California

106. Which city is farther south – Indianapolis, Houston, or Detroit?
Houston

107. Which river forms the eastern boundary of Missouri – Mississippi or Arkansas?
Mississippi

108. The nuclear-powered U.S.S. Virginia was built at Groton, home of the U.S. Naval Submarine Base in what northeastern state?
Connecticut

109. Each year, contestants bring their pumpkins and launching machines to the Punkin Chunkin World Championship in Bridgeville, a city in what state whose most populous city is Wilmington?
Delaware

110. What state is the leading harvester of wild blueberries in the United States?
Maine

111. What mountain in New Hampshire holds the record for the highest surface wind speed?
Mount Washington

112. Atlantic City and Newark are cities in what state?
New Jersey

113. Cooperstown is the home of the National Baseball Hall of Fame in what state with the cities of Syracuse and Amherst?
New York

114. Narragansett Bay can be found in what state where East Providence is located?
Rhode Island

115. What state bordering New Hampshire is the United States' leading producer of maple syrup?
Vermont

116. Theodore Roosevelt National Park can be found in what state with Spirit Lake Dakotah Nation?
North Dakota

117. Taum Shak Mountain can be found in what state home to the Lake of the Ozarks?

Missouri

118. Lake Pontchartrain borders what state bordering Atchafalaya Bay?
Louisiana

119. President Ronald Reagan was born in what state with the cities of Aurora and Rockford and is also home to Shawnee National Forest?
Illinois

120. Since the 1820s, a plant called leafy spurge, which is difficult for ranchers to control, has spread, causing problems in what grassland region in the United States?
Great Plains

121. What valley lies between the Allegheny Mountains and the Blue Ridge Mountains?
Shenandoah Valley

122. In what U.S. state could you play midnight baseball in June without artificial lighting?
Alaska

123. Richard B. Russell Lake can be found between South Carolina and what other state?
Georgia

124. Monongahela National Forest can be found in what state with the cities of Summersville and Fayetteville?
West Virginia

125. Missouri National Recreational River comprises part of the border between Nebraska and what other state whose major natural sites include Sioux Falls and Black Hills National Forest?
South Dakota

126. Philippine languages, Chinese, Chamorro, and English are the main languages of what U.S. territory in the Pacific Ocean?
The Northern Mariana Islands

127. Casper is a city in what state where the John D. Rockefeller Memorial Parkway can be found?
Wyoming

128. Puget Sound, which lies between the Olympic Mountains and the Cascade Range, is a major inlet of which ocean?
Pacific Ocean

129. The Adirondack Mountains, which cover much of the area between Lake Ontario and Lake Champlain, are in the northern part of which state?
New York

130. Which river forms the southeastern boundary of Illinois?
Ohio River

131. Which state is closest to Russia?
Alaska

132. Which state has more wildlife refuges – Oregon or Kentucky?
Oregon

133. Which state borders Lake Champlain – Vermont or New Hampshire?
Vermont

134. Which state has a larger population – Connecticut or West Virginia?
Connecticut

135. The Mississippi River separates Tennessee from which state—Arkansas or Oklahoma?
Arkansas

136. Which state is north of the Ohio River—Virginia or Indiana?
Indiana

137. Which state, known as the Centennial State, joined the Union in 1876—Colorado or Vermont?
Colorado

138. Which river forms most of the border between Georgia and South Carolina—the Potomac River or the Savannah River?
Savannah River

139. Which city lies near the junction of the Missouri and Mississippi Rivers—Memphis or St. Louis?
St. Louis

140. Which agricultural state is the geographical center of North America—New Jersey or North Dakota?
North Dakota

141. Farmers in which state use water from the Ogallala Aquifer for irrigation—Nebraska or Maine?
Nebraska

142. Which city is located at the junction of the Allegheny and Monongahela Rivers—Minneapolis or Pittsburgh?
Pittsburgh

143. Which city is Oregon's largest city—Portland or Helena?
Portland

144. The Sea Island chain, including Hilton Head and Cumberland Island, runs along the coast of which state—Michigan or South Carolina?
South Carolina

145. Which city is located on the Cumberland River and is known for its many country music recording centers—Nashville or Little Rock?
Nashville

146. Which city is known for its extensive freeway system and is located in the most populous county in the United States—Dallas or Los Angeles?
Los Angeles

147. Which city was founded as a fort between Lake Erie and Lake Huron and developed into one of the country's leading manufacturing centers—Detroit or Milwaukee?
Detroit

148. Which city is located on a river delta—New Orleans or Austin?
New Orleans

149. Which city was named after the founder of a famous stagecoach line that once crisscrossed the West—Fargo or Boston?
Fargo

150. Which city is the westernmost point on the Erie Canal—Cincinnati or Buffalo?
Buffalo

151. Which major city is located in North Carolina and was near the site of the first important gold discovery in the United States—Columbia or Charlotte?
Charlotte

152. The world's largest naval base is located at the mouth of the Chesapeake Bay in which city—Norfolk or Washington, D.C.?
Norfolk

153. Which city is located on the western tip of Lake Superior and is an important port for shipping grain and iron ore—Des Moines or Duluth?

Duluth

154. Which city was badly damaged during an earthquake in 1964 and is now the most populous city in Alaska—Anchorage or Barrow?
Anchorage

155. Which state is not a major producer of wheat—Connecticut, North Dakota, or Montana?
Connecticut

156. Which of the following states is not crossed by the Colorado River—Arizona, Colorado, or New Mexico?
New Mexico

157. Which state does not border Tennessee—North Carolina, Mississippi, or West Virginia?
Virginia

158. Which of the following is not a Great Plains state—Ohio, Oklahoma, or Kansas?
Ohio

159. Which state's capital is not named after a President of the United States—Mississippi, Nebraska, or South Dakota?
South Dakota

160. Which state does not include part of Yellowstone National Park—Idaho, Oregon, or Wyoming?
Oregon

161. Which state does not have a continental climate—Iowa, Minnesota, or Mississippi?
Mississippi

162. Which state does not border Saskatchewan—Montana, North Dakota, or Wisconsin?
Wisconsin

163. Which state is not part of New England—Pennsylvania, New Hampshire, or Connecticut?
Pennsylvania

164. Which state does not have a panhandle—Colorado, Florida, or Oklahoma?
Colorado

165. Which state does not experience hurricanes – North Carolina, Iowa, or Florida?
Iowa

166. Which state does not have a large Hispanic population – Texas, Montana, or Arizona?
Montana

167. Badlands National Park and Mount Rushmore are west of the Missouri River in which state?
South Dakota

168. Which city lies along the Willamette River – Portland or Sacramento?
Portland

169. Most of Yellowstone National Park is in which state?
Wyoming

170. Kodiak, Nunivak, and Baranof are islands that are part of which state?
Alaska

171. Which New England city is located north of Boston and witnessed the execution of alleged witches in 1692?
Salem

172. Which body of water is partially surrounded by Maryland?
Chesapeake Bay

173. The aggressive Formosan termite has invaded southern cities, damaging wooden homes and buildings. This infestation is particularly destructive in which Louisiana city known for its French and Cajun traditions?
New Orleans

174. Lake Powell, a reservoir on the Colorado River, is shared by Arizona and which state?
Utah

175. "Virtue, Liberty, and Independence" is the motto of which middle Atlantic state that is home to Valley Forge National Historical Park?
Pennsylvania

176. Which state is not bordered by a major body of water – Michigan, California, or Wyoming?
Wyoming

177. Which state has the lowest per capita income – Texas or Mississippi?
Mississippi

178. Which is the largest U.S. lake west of the Mississippi River – The Great Salt Lake or Lake Mead?
The Great Salt Lake

179. Which state has a larger Native American population – Oklahoma or Maryland?
Oklahoma

180. Washington, D.C., was created from portions of which two states?
Virginia and Maryland

181. Which state reaches farther north – Wyoming or West Virginia?
Wyoming

182. Which state has more people per square mile – Connecticut or Kentucky?
Connecticut

183. The Great Basin can be found in which state – Nevada or Utah?
Nevada

184. Which of the following states was not on the Oregon Trail – Wyoming, Tennessee, or Nebraska?
Tennessee

185. The Brooks Range in Alaska acts as a divide for waters flowing north into the Arctic Ocean and south into what river that flows across the state into the Bering Sea?
Yukon River

186. Which state borders Lake Erie – Pennsylvania or Illinois?
Pennsylvania

187. Which crop is more important to the economy of Florida – Oranges or Wheat?
Oranges

188. Which of the following states was not part of the Cotton Belt – South Carolina, Maryland, or Georgia?
Maryland

189. The name of which state comes from the French words for "Green Mountain"?
Vermont

190. Louisville, located on the Ohio River, is a major city in which state?
Kentucky

191. What type of climate do you find in the Great Plains – tropical or semi-arid?

Semi-arid

192. Which type of climate is found along the coast of Oregon – Mediterranean or marine west coast?
Marine west coast

193. Which city is located in the eastern part of Washington State – Spokane or Auburn?
Spokane

194. The Rogue River runs through Grants Pass as it flows from the Cascades to the Pacific Ocean in which state?
Oregon

195. The Shoshone Falls are near the city of Twin Falls in which U.S. state?
Idaho

196. Roanoke is in what present-day state bordering Chesapeake Bay – Virginia or Maryland?
Virginia

197. Long Island Sound borders what state to the north – New York or Connecticut?
Connecticut

198. Block Island is a famous island belonging to what state with the capital city of Providence – Rhode Island or Vermont?
Rhode Island

199. Ben & Jerry's can be found in what state known as the Green Mountain State – Vermont or New Hampshire?
Vermont

200. Which city is located on the northeastern shore of San Francisco Bay?
Oakland

201. Which state has a lower average elevation – Florida or Arkansas?
Florida

202. The Alamo attracts many visitors to San Antonio, a city in which state?
Texas

203. Which city is located at a higher elevation – Knoxville or Kansas City?
Knoxville

204. "With God, All Things Are Possible" is the motto of which midwestern state that is home to Cuyahoga Valley National Park?
Ohio

205. Which state has a greater gross state product – Kentucky or New York?
New York

206. Which city is the westernmost point on the Erie Canal – Syracuse or Buffalo?

Buffalo

207. Which state does not have a continental climate – Mississippi, Minnesota, or Iowa?
Mississippi

208. Which state is not part of New England – Pennsylvania, New Hampshire, or Connecticut?
Pennsylvania

209. "Agriculture and Commerce" is the motto of what southern state that is home to much of the Cumberland Plateau?
Tennessee

210. Which river runs between Arizona and California before entering Mexico – Platte River or Colorado River?
Colorado River

211. The Cherokee Native Americans were forced to relocate to present-day Oklahoma from what region of the United States?
Southeast

212. Which state borders more Great Lakes – Michigan or New York?
Michigan

213. The Merrimack River, formed by the junction of the Pemigewasset and Winnipesaukee Rivers, empties into what major body of water?

Atlantic Ocean

214. Which state is adjacent to the Gulf of Mexico – Mississippi or Georgia?
Mississippi

215. Which city is located in an arid climate zone – Santa Fe or Milwaukee?
Santa Fe

216. Which city is famous for its steep hills and cable cars – San Francisco or San Antonio?
San Francisco

217. Which state includes Rocky Mountain National Park—Kansas or Colorado?
Colorado

218. The San Andres Mountains parallel the Rio Grande as it flows through which state?
New Mexico

219. Which northeast state borders Canada – New Hampshire, New Jersey, or Massachusetts?
New Hampshire

220. Mauna Loa is a large shield volcano located in what state?
Hawaii

221. The geographic center of the lower 48 states lies in what state?

The Geography Bee Ultimate Preparation Guide

Kansas

222. The wettest place in the United States is Mount Waialeale in what state?
Hawaii

223. Which state is more likely to experience a tornado – Utah or Oklahoma?
Oklahoma

224. Which state capital is closer to the Gulf of Mexico – Little Rock or Olympia?
Little Rock

225. What state is home to the U.S. Navy's Fleet Forces Command?
Virginia

226. Galveston is to Texas as Pensacola is to what state?
Florida

227. Havre is in the northern region of what state, with the cities of Billings and Missoula?
Montana

228. The Columbia River forms part of the border between Oregon and what other state?
Washington

229. Nebraska borders Kansas, Colorado, Wyoming, South Dakota, Iowa, and what other state to the southeast?

Missouri

230. Barrow and Wainwright are cities on the northern coast of what U.S. state?
Alaska

231. What sport played widely in the United States is known as "America's National Pastime"?
Baseball

232. The Kennedy Space Center is located on the coast of what U.S. state?
Florida

233. The city of Alturas can be found in the northern region of what western U.S. state?
California

234. The Colorado River carves out a famous canyon in the southwestern United States. Name this canyon.
The Grand Canyon

235. The Upper and Lower Peninsulas of Michigan border two of the Great Lakes. One of these lakes is Lake Michigan. What is the other lake?
Lake Huron

236. Atchafalaya Bay, bordering the Gulf of Mexico, can be found south of what state?
Louisiana

The Geography Bee Ultimate Preparation Guide

237. Rapid City is located northeast of Mount Rushmore National Monument in what state?
South Dakota

238. Lāna'i City is located on the island of Lānai in what U.S. state?
Hawaii

239. Laredo and Brownsville are cities close to the U.S.'s border with Mexico in what state?
Texas

240. The Seward Peninsula in Alaska is east of the Bering Sea and north of what sound?
Norton Sound

241. What tree is the national tree of the United States?
Oak Tree

242. Hyperion, the world's tallest living tree, can be found in Redwood National Forest. This tree can be found in what state?
California

243. The Chrysler Building can be found in what northeastern U.S. state?
New York

244. Castle Geyser can be found in what famous U.S. national park?
Yellowstone National Park

245. What state is known as the "lightning capital of the United States"?
Florida

246. You can ride a train down Candy Alley at the Jelly Belly warehouse near Kenosha, a city located on Lake Michigan in which state – Wisconsin or Indiana?
Wisconsin

247. You can create your own custom PEZ candy dispenser on the PEZ factory tour near New Haven in which New England state – Maine or Connecticut?
Connecticut

248. What city is known as the "Mile-high city"?
Denver

249. Which river is farther north – Potomac River, Rio Grande, or Chattahoochee River?
Potomac River

250. The Grand Canyon is located in what U.S. state?
Arizona

251. Lake Okeechobee is a physical landmark in what southeastern state?
Florida

252. The Chinese tallow tree, whose seeds can be used in candles and soap, has spread throughout the southeastern

United States. This tree was first introduced to what state located northeast of Georgia?
South Carolina

253. Which U.S. state has a shape that includes a panhandle – Wyoming, Oklahoma, or Colorado?
Oklahoma

254. To see giant sequoia trees and the Channel Islands, you would travel to which U.S. state – California or Virginia?
California

255. What city, founded by the Dutch near the mouth of the Hudson River, was once known as New Amsterdam?
New York City

256. Most of the wheat in the United States is grown in which region – Great Plains or Columbia River Valley?
Great Plains

257. Laramie is a city in what landlocked state bordering Colorado to the south?
Wyoming

258. Kotzebue is a city on the coast of what state?
Alaska

259. The San Juan Mountains can be found in the southern region of what state?
Colorado

260. Yuma is near the border with California in what state?
Arizona

261. Pensacola is located in the western part of what state?
Florida

262. Cape Hatteras is a famous cape located in what state with the city of Greensboro?
North Carolina

263. The Upper and Lower Peninsulas belong to what state?
Michigan

264. Alturas is located in the northeastern region of what state?
California

265. Fort Worth is west of what major Texan city?
Dallas

266. Galveston borders what body of water?
Gulf of Mexico

267. Atchafalaya Bay is located south of what state?
Louisiana

268. Vermont and New Hampshire border what Canadian province?
Quebec

269. The Kennedy Space Center is located on what cape?
Cape Canaveral

270. Norton Sound, south of the Seward Peninsula, is an inlet of what strait?
Bering Strait

271. The Great Basin can be found primarily in what state?
Nevada

272. Cape Cod borders what major gulf bordering Maine?
Gulf of Maine

273. New Haven is a major city in what state where the Connecticut River can be found?
Connecticut

The Geography Bee Ultimate Preparation Guide

North America

1. Parliament Hill, site of Canadian national government buildings, is in which city – Ottawa or Montreal?
   Ottawa

2. The Lempa River, which flows into the Pacific Ocean, provides hydroelectric power in which Central American country – El Salvador or Guatemala?
   El Salvador

3. Prince Rupert's Land is a historic region that made up much of which present-day country in the Western Hemisphere – Canada or Argentina?
   Canada

4. Stromboli is to Italy as Popocatépetl is to what – Mexico or Guatemala?
   Mexico

5. What country has the third largest area in the world – Mexico or United States?
   United States

6. Which of these island countries has a smaller total land area – Cuba, Bahamas, or Jamaica?
   Bahamas

7. The ports of Anchorage, Vancouver, and Acapulco are along which coast of North America – West or East?
West

8. Hispaniola consists of the Dominican Republic and what other country – Jamaica or Haiti?
Haiti

9. Curacao is the largest island in what Antilles – Netherlands Antilles or Lesser Antilles?
Netherlands Antilles

10. Saint Barthelemy is located in what North American region – the Caribbean or Central America?
Caribbean

11. Aguascalientes is a province in what country – Mexico or Cuba?
Mexico

12. Prince Edward Island is the smallest province in what country – Canada or the Bahamas?
Canada

13. The Yucatan Peninsula is located on what country – Panama or Mexico?
Mexico

14. The Denmark Strait separates Iceland and what other territory – Greenland or Bermuda?

Greenland

15. The Turks and Caicos Islands belong to what European country – United Kingdom or France?
United Kingdom

16. The Statue of Liberty, a gift from France, is located in what U.S. State – New York or California?
New York

17. Aruba, Bonaire, and Curacao are part of what region of the Caribbean – Lesser Antilles or Netherlands Antilles?
Netherlands Antilles

18. Oranjestad is the capital of what Dutch Territory in the Caribbean – Bonaire or Aruba?
Aruba

19. St. Pierre and Miquelon is a territory of the United Kingdom nearest to what country – Canada or the United States?
Canada

20. Ellesmere Island and Baffin Island are two large islands in what country – Mexico or Canada?
Canada

21. Godthab is another name for the capital of Greenland, which is what – Nuuk or Resolute?
Nuuk

The Geography Bee Ultimate Preparation Guide

22. The Beaufort Sea is located north of Canada and a U.S. state. Name this state – Hawaii or Alaska?
Alaska

23. Kingstown is the capital of what country located in the Caribbean that gained independence from the United Kingdom in 1979 – St. Vincent and the Grenadines or St. Kitts and Nevis?
St. Vincent and the Grenadines

24. The Sierra Madre Occidental and the Sierra Madre Oriental are two mountain ranges in what country – Mexico or United States?
Mexico

25. The Yucatan Channel separates what country from Mexico?
Cuba

26. Andros and Great Inagua are the two largest islands of what country whose highest point is Mount Alvernia and is a popular tourist destination south of the U.S. state of Florida?
Bahamas

27. Known for its Blue Mountains, which country is the third largest island in the Caribbean?
Jamaica

28. The Maya Mountain Range, which extends into Guatemala, can also be found in what country that borders the Bay of Chetumal, Gulf of Honduras, and Caribbean Sea?

Belize

29. Gaston Browne is the current prime minister of what country in the Caribbean that is predominantly Protestant and whose capital is Saint John's?
Antigua and Barbuda

30. Mount Hillaby is the highest point in what Caribbean country whose capital is Bridgetown?
Barbados

31. The Gulf of Fonseca forms the northwestern coast of what country whose second largest lake is Lake Managua?
Nicaragua

32. Which island country of the Greater Antilles uses English as its official language?
Jamaica

33. What river forms part of the boundary between Costa Rica and Nicaragua – San Juan or Rio Grande?
San Juan

34. The Great Northern Peninsula, which extends approximately 170 miles from Bonne Bay on the west to Cape Bauld at the tip, is located in Newfoundland in which country – Canada or the United States?
Canada

35. Lake Gatun, an artificial lake that constitutes part of the Panama Canal system, was created by damming which river – Chagres River or Amazon River?
Chagres River

36. What large island is located 200 miles northwest of Iceland – Baffin Island or Greenland?
Greenland

37. The largest metropolitan area in North America, is also the capital of a country. Name this metropolitan area – Washington D.C. or Mexico City?
Mexico City

38. What is the most densely populated country in North America, located in the Caribbean with the capital of Bridgetown – Trinidad and Tobago or Barbados?
Barbados

39. What valley is the lowest point in North America, located near the Mojave Desert?
Death Valley

40. The Anguilla Cays, part of the Bahamas, lie north of the Nicholas Channel about 50 miles from which island?
Cuba

41. Since 1960, thousands of people have migrated across the Straits of Florida to the United States from which country?
Cuba

The Geography Bee Ultimate Preparation Guide

42. Which island country is not part of the Greater Antilles – Barbados, Jamaica, or Cuba?
    Barbados

43. Which of these islands is NOT in the Caribbean Sea – Bermuda, Puerto Rico, or Hispaniola?
    Bermuda

44. In the summer of 2013, astronauts captured images of a flooding in a Canadian city located at the confluence of the Bow and Elbow Rivers. Name this city.
    Calgary

45. The wreck of the *Titanic* was discovered in 1985, a few hundred miles off the coast of Cape Race, which is located on what peninsula?
    Avalon Peninsula

46. A Mexican state that borders Arizona shares its name with a desert. Name this Mexican state.
    Sonora

47. Chiles en Nogada, a dish of meat-filled peppers in walnut cream sauce, represents the colors of the Mexican flag and is popular in the country's fourth-largest city. Name this state capital city located east of Popocatepetl?
    Puebla

48. Activity along the San Andreas Fault in North America helped create which peninsula – Baja California or Yucatan Peninsula?

Baja California

49. The largest city in northern Haiti was renamed following Haiti's independence from France. What is the present-day name of this city?
Cap-Haitien

50. The Churchill, Slave, and Peace Rivers are in what country?
Canada

51. The so-called Barren grounds east of the Mackenzie River basin is a tundra region in which country?
Canada

52. What Central American country was once a province within Colombia?
Panama

53. Baja California, which is almost completely surrounded by water, is an example of what kind of physical feature?
Peninsula

54. Western Canada's most populous metropolitan area has a large number of Chinese immigrants. Name this city.
Vancouver

55. Baja California and Yucatan are peninsulas of what country?
Mexico

56. Which city, sometimes called the Gateway to the West, is near the junction of the Mississippi and Missouri Rivers?

St. Louis

57. Mount McKinley, the highest peak in North America, is in what U.S. state?
Alaska

58. Ciudad Juarez and Tijuana are both major cities in what country bordering the Gulf of Mexico and Guatemala?
Mexico

59. Which of these countries is part of Central America – Guyana, Costa Rica, or Uruguay?
Costa Rica

60. Mt. Liamuiga, at 3,793 feet, is the highest peak in what country whose main natural hazard is hurricanes, produces sugarcane and rice, and is home to the Great Salt Pond?
St. Kitts and Nevis

61. What country, which gained its independence from Spain, has a tropical climate and contains the Guantanamo Naval Base?
Cuba

62. Jiquilisco Bay is in what country whose geographical features include the Cerrón Grande Reservoir and Lake Ilopango?
El Salvador

63. Which of these Caribbean islands is closest to the Tropic of Cancer – Cuba or Grenada?

Cuba

64. Sri Lanka is to island as Panama is to what?
Isthmus

65. The Sonoran Desert, in northwestern Mexico, is west of what major Mexican mountain range covering much of western Mexico?
Sierra Madre Occidental

66. Kerala is to India as San Luis Potosi is to what?
Mexico

67. Which correctly describes the location of Edmonton, Alberta – on the Mackenzie River or north of Calgary?
North of Calgary

68. Central America's most populous country has more than 12 million people. Name this country, which borders Mexico.
Guatemala

69. What North American river connects the Great Lakes with the Atlantic Ocean?
St. Lawrence River

70. To visit writer Ernest Hemingway's home near Havana and to hike to the top of El Yunque, you would travel to what country?
Cuba

71. The country with the longest coastline in the world is located on which continent?
North America

72. Windsor lies at the western edge of Canada's most heavily populated area. This city is in what province?
Ontario

73. The Panama Canal links the Pacific Ocean with which other body of water?
Caribbean Sea

74. The Gulf Stream is a warm ocean current in what ocean?
Atlantic Ocean

75. Whistler and Blackcomb are mountains in what country that hosted the 2010 Winter Olympics?
Canada

76. Kangiqsujuaq is a city on the Hudson Strait in what country?
Canada

77. The Arch is located at the tip of Baja California in what country?
Mexico

78. Teotihuacan was a holy city with pyramids along the Avenue of the Dead in what country?
Mexico

79. Mexico borders the United States, Belize, and what other country to the southeast?
Guatemala

80. Tikal National Park and Lake Izabal are in what country bordering Belize?
Guatemala

81. Atitlan Nature reserve can be found in the western region of what country?
Guatemala

82. The Northern Lagoon and Southern Lagoon can be found in what country bordering the Gulf of Honduras?
Belize

83. Blue Hole National Monument is east of the Turneffe Islands in what country?
Belize

84. Nueva San Salvador is west of San Salvador, a city in what country bordering the Gulf of Fonseca?
El Salvador

85. Lago de Yojoa and the Cruta River can be found in what country?
Honduras

86. Estadio Cuscatlan is the largest stadium in Central America. This famous stadium is in what small country with an excellent national soccer team?

El Salvador

87. Bosawas Biosphere Reserve is located in what country where Lake Managua is found?
Nicaragua

88. Masaya Volcano is the most active volcano in what country that is home to the cities of Masaya and Chinandega?
Nicaragua

89. Monteverde Forest Reserve can be found in what country where the Osa and Nicoya Peninsulas can be found?
Costa Rica

90. Mount Chirripo is the highest mountain in what country bordering Coronado Bay and Dulce Gulf?
Costa Rica

91. Amador Causeway is a road starting at a famous canal, and it leads to four islands. This road is located in what country known as the "Crossroads of the World"?
Panama

92. You can tour coffee plantations in Boquete in what country with the cities of David and La Chorrera?
Panama

93. Dunn's River Falls attracts many tourists in what country famous for its many athletes?
Jamaica

94. Valley of the Sugar Mills is located in what country bordering the Nicholas Channel?
Cuba

95. Castillo de la Real Fuerza is Havana's oldest fort, built to protect against pirates. This fort is located in what country?
Cuba

96. Quisqueya Stadium is located in Santo Domingo, in what country?
Dominican Republic

97. The Marinarium can be found in Punta Cana, in what country bordering Manzanillo Bay?
Dominican Republic

98. The Citadelle is a fortress in what French-speaking Caribbean country in the Greater Antilles?
Haiti

99. Eleuthera Island and Andros Island can be found in what country bordering the Northwest Providence Channel and the Mayaguana Passage?
The Bahamas

100. Brimstone Hill Fortress National Park can be found in Sandy Point Town in what tiny country?
St. Kitts and Nevis

101. The U.S. Virgin Islands can be found south of what British territory?

British Virgin Islands

102. What Caribbean overseas territory of France has Fort-de-France as its capital and Le Lamentin as one of its major cities?
Martinique

103. Christiansted and Frederiksted are cities on St. Croix in what Caribbean territory of the United States?
U.S. Virgin Islands

104. Barbados is located nearly 100 miles east of what country just north of Grenada?
St. Vincent and the Grenadines

105. Hindus in the Caribbean are found mainly in what country bordering Serpent's Mouth and the Gulf of Paria?
Trinidad and Tobago

106. The Labrador Sea is an arm of what ocean?
Atlantic Ocean

107. The Strait of Canso separates mainland Canada from what island?
Cape Breton Island

108. Cape Fear borders what body of water?
Atlantic Ocean

109. The Tropic of Cancer passes through which country – Mexico, Canada, or Brazil?

Mexico

110. The island of Newfoundland is located off the east coast of Canada in which ocean – Atlantic Ocean or Pacific Ocean?
Atlantic Ocean

111. What is the name of the ocean current that helps create rich fishing grounds in the western part of the Atlantic Ocean?
Labrador Current

112. Golfe de la Gonave is a gulf forming the west coast of what French-speaking Caribbean country?
Haiti

113. What island country borders Mayaro Bay, Cocos Bay, and Saline Bay?
Trinidad and Tobago

114. Which city is in western Texas and lies directly across the border from the Mexican city of Ciudad Juarez?
El Paso

115. If you were in Chicago, Illinois, which would you be closest to – the Tropic of Cancer or the Tropic of Capricorn?
Tropic of Cancer

116. Alberta and which other Canadian province are landlocked?
Saskatchewan

The Geography Bee Ultimate Preparation Guide

South America

1. Montevideo is a major port on the northern shore of the Rio de la Plata. Name the country in which it is located – Uruguay or Paraguay?
   Uruguay

2. The Chaco, a semiarid plain that is home mostly to cattle ranchers, makes up most of the western half of what landlocked country in South America – Paraguay or Bolivia?
   Bolivia

3. Pamperos are strong wind and rainstorms that characterize the Pampas and the northeastern region of what South American country – Venezuela or Argentina?
   Argentina

4. If you were visiting Rio de Janeiro, which language would you expect most residents of the city to speak – Portuguese or Spanish?
   Portuguese

5. Which Colombian city lies on a plateau just west of the Llanos plains, a grassland region that stretches through eastern Colombia and western Venezuela – Bogota or Barranquilla?
   Bogota

6. What is another name for the Falkland Islands, a territory of the United Kingdom located east of Argentina – Islas Malvinas or the British Isles?
Islas Malvinas

7. Paraguay and what other country are the only landlocked countries in South America – Uruguay or Bolivia?
Bolivia

8. French Guinea is an overseas department of France, bordering Brazil to the south and what country to the west – Guyana or Suriname?
Suriname

9. Easter Island belongs to what country – Peru or Chile?
Chile

10. The Equator passes through what country that is home to Chimborazo – Ecuador or Brazil?
Ecuador

11. Panama borders what country – Venezuela or Colombia?
Colombia

12. Ushuaia is regarded as the southernmost city in the world, located on Terra del Fuego, an island shared by Chile and what other country – Bolivia or Argentina?
Argentina

13. Christianity and what other religion are the two largest in Suriname and Guyana – Hinduism or Islam?
Hinduism

14. The Galapagos Islands belong to what country – Paraguay or Ecuador?
Ecuador

15. What country has an area almost half the size of South America – Argentina or Brazil?
Brazil

16. Paraguay borders Brazil, Argentina, and what other country to the northwest that does not have a coastline – Peru or Bolivia?
Bolivia

17. Machu Picchu is an Inca site located in the Urubamba province of what country bordering the Pacific Ocean – Peru or Colombia?
Peru

18. The Amazon River is connected to what ocean – Indian Ocean or Atlantic Ocean?
Atlantic Ocean

19. La Paz and Cochabamba are major cities of what country with the plateau region of Altiplano in the southwest?
Bolivia

20. Patagonia is a region in the southern part of what country whose major rivers are the Paraguay, Parana, and Uruguay rivers?
Argentina

21. The Amazon and the Paraná Rivers both flow through what country?
Brazil

22. Tungurahua is an active volcano in what country whose most populous city is Guayaquil?
Ecuador

23. Angel Falls is to Venezuela as Three Sisters Falls is to what country?
Peru

24. The driest place in the world is in which South American desert?
Atacama Desert

25. Which country does not include a portion of the Andes Mountains – Brazil, Peru, or Argentina?
Brazil

26. A cold ocean current that flows along the southeastern coast of South America shares its name with a group of nearby islands. Name this ocean current.
Falkland Current

27. Which of the following South American capital cities is not on the banks of the Rio de la Plata – Montevideo, Buenos Aires, or Brasilia?
Brasilia

28. Name the flat intermontane area located at an elevation of about 10,000 feet (3,050 m) in the central Andes.
Altiplano

29. The Llanos is a grassland region that extends from Colombia into what other country?
Venezuela

30. What capital city located on the Suriname River has distinctive Dutch colonial architecture?
Paramaribo

31. Cochabamba is the third largest urban area of what country?
Bolivia

32. Because Earth bulges at the Equator, the point that is farthest from Earth's center is the summit of a peak in Ecuador. Name this peak.
Chimborazo

33. San Agustín was an important ceremonial center in what country north of Ecuador?
Colombia

34. Which river is not located in South America – Amazon, Mekong, or Orinoco?
Mekong

35. Los Llanos is a large tropical grassland plain in Venezuela and what other country straddling the Sierra Nevada de Santa Marta, the world's highest coastal range – Colombia or Guyana?
Colombia

36. Which country has a drier climate - Argentina or Colombia?
Argentina

37. Wonotobo Vallen, a lake on the edge of the Guiana Highlands, is in what country whose capital is a chief port on the Atlantic Ocean?
Suriname

38. Which country is not bordered by the Orinoco River – Colombia or Ecuador?
Ecuador

39. Name the highlands in northern South America that extend across much of the area between the Orinoco and Amazon river basins.
Guiana Highlands

40. The Itaipu Reservoir, on the border between Paraguay and Brazil, is fed by what major South American River?
Parana River

The Geography Bee Ultimate Preparation Guide

41. Mt. Huascaran, at 22,205 feet (6768 meters) is the highest point in what country?
Peru

42. Bolivia and what other South American country is landlocked?
Paraguay

43. Rio de Janeiro and Sao Paulo are major cities in what South American country?
Brazil

44. In March 1998 fires burned a large area near the rain forest home of the Yanomami people. These fires occurred in the northern part of which country in the Western Hemisphere?
Brazil

45. Pampas is a term used for temperate grasslands on which continent?
South America

46. The archipelago Islas Malvinas, also called the Falkland Islands, lies off the southeastern edge of which continent?
South America

47. On which continent would you find the longest mountain system above sea level and the southernmost city in the world?
South America

48. The urban area of Cochabamba has been in the news in recent years due to protests over the privatization of the municipal water supply and regional autonomy issues. Cochabamba is the third largest conurbation in what country?
Bolivia

49. After an economic collapse in the late 1990s, what South American country declared the U.S. dollar to be its official currency, replacing the sucre?
Ecuador

50. Which South American country was a member of OPEC until 1992 – Ecuador or Chile?
Ecuador

51. Lima is home to more than a quarter of the population of a country on the Pacific Ocean. Name the country in which it is located.
Peru

52. Medellin, known for hosting the biggest flower parade in the world, is located in what country?
Colombia

53. You can tour a banana plantation and learn how they grow from tiny red flowers to the fruit we eat in Guayaquil, in what country?
Ecuador

54. Turtumo Volcano is a major tourist attraction in what city near the coast of Colombia?
Santa Catalina

55. 36 islands make up Mochima National Park and you can swim and snorkel in its coral reefs. This national park is located in what country home to the cities of Guanare and Paraguaipoa?
Venezuela

56. The Hindu holiday of Phagwah, also known as Holi, is celebrated actively in Georgetown. Georgetown is the capital of what country?
Guyana

57. Shell Beach can be found on the northern coast of what country where the Barima and Cuyuni Rivers can be found?
Guyana

58. The Wilhelmina Mountains can be found in what country that claims land in French Guiana and Guyana?
Suriname

59. One of the most powerful waterfalls lies in Guyana's rainforest. At 741 feet, what is the name of this famous waterfall?
Kaieteur Falls

60. Corcovado Mountain is located in a major city in southeastern Brazil. Name this city.
Rio de Janeiro

61. Giant anteaters and otters can be found in what floodplain in Brazil?
Pantanal

62. Morumbi Stadium, with 84,000 seats, is the largest in what country?
Brazil

63. Lake Titicaca is shared between Bolivia and what other country home to the cone-shaped peak of Volcan Misti?
Peru

64. Pilpintuwasi Butterfly Farm can be found in Iquitos, a city in the northern region of what country?
Peru

65. Isla del Sol is an island in what major high navigable lake in Bolivia and Peru?
Lake Titicaca

66. San Pedro Hill is a popular tourist attraction in Cochabamba. Cochabamba lies near the Cordillera Oriental mountain range in what country?
Bolivia

67. Uyuni Salt Flat is the world's largest inland area of salt, in what country where the world's highest capital can be found?
Bolivia

68. Vina del Mar, known as the "Garden City", is located very close to Valparaiso in what country?
Chile

69. The Chonos Archipelago is located in what country home to Torres del Paine National Park and Malalcahuello-Nalcas National Reserve?
Chile

70. Iguazu Falls is a set of two waterfalls in what country whose currency is the peso?
Argentina

71. Jaguars and pumas roam Defensores del Chaco National Park. Here, you can spot tapirs and colorful birds as well. This national park is located in the northwest region of what country?
Paraguay

72. The Museum of Clay is located in the capital of Paraguay, near the border with Argentina. Name this capital city.
Asuncion

73. Lake Rincón del Bonete is located in what country that is home to the Cuchilla Grande and the beaches of Punta Del Este?
Uruguay

74. The Orinoco, Casiquiare, and Arauca Rivers are located in what country bordering the Guajira Peninsula in Colombia?
Venezuela

75. South America's lowest elevation is found on the Valdés Peninsula. This peninsula is along the Atlantic coast of what country?
Argentina

76. Estadio Centenario is a stadium in Uruguay, built in 1930 for the world's first World Cup. This stadium is for what popular South American sport?
Soccer

77. Inti Raymi is celebrated by Amerindians and is known commonly by what name?
Festival of the Sun

78. Buenaventura Bay borders Ecuador and what other country?
Colombia

79. The Rio de la Plata, bordering Uruguay and Argentina, empties out into what ocean?
Atlantic Ocean

80. São Luis borders São Marco Bay in what country?
Brazil

81. Bogota and Medellin are cities in which South American country?
Colombia

82. The Drake Passage lies south of what continent?

South America

83. The Patagonia Region covers much of the southern part of which continent – South America or Africa?
South America

84. Brazil and Uruguay border what ocean?
Atlantic Ocean

85. The Andes Mountains are located on what continent?
South America

86. Pato, a sport originally played by the gauchos, is popular in which country?
Argentina

87. The Itiapu Reservoir is fed by what river?
Parana River

88. Galera Point is located in what country bordering Buenaventura Bay?
Ecuador

89. Blanca Bay borders the city of Bahia Blanca in what country?
Argentina

90. Cape Santa Marta Grande is located miles away from Florianopolis in what country?
Brazil

91. Fortaleza is a major city in what country where the Sobradinho Reservoir can be found?
Brazil

92. Gorgona Island belongs to what country bordering the Gulf of Uraba?
Colombia

93. Loja and Machala are major cities in the southern region of what country that straddles the Equator in the north?
Ecuador

94. Oil and gas make what country one of South America's richest countries?
Venezuela

95. You can find capybaras at Hato el Cedral in what country?
Venezuela

96. What country is home to Lake Rogagua and Lake San Luis?
Bolivia

97. San Matias Gulf is a major gulf bordering the Valdes Peninsula in what country?
Argentina

98. The Wollaston Islands belong to what country bordering the Strait of Magellan?
Chile

99. The Kwitaro and Essequibo rivers are located in what country straddling the Pakaraima Mountains?
Guyana

100. Brownsberg Nature Park is located in what country, home to the Tapanahoni River?
Suriname

Asia

1. Which Indian state has the highest literacy rate in India – Jharkhand or Kerala?
   Kerala

2. Which country is the most populous in Asia – China or Indonesia?
   China

3. Which South Asian country, the eighth most populous in the world, was previously known as East Pakistan – Bangladesh or Myanmar?
   Bangladesh

4. The Kizilirmak River is also known as what river – Huang He or Halys River?
   Halys River

5. The Ural Mountain Range divides Russia into Asia and what other continent – North America or Europe?
   Europe

6. Zoroastrianism began in what Middle Eastern country that borders Afghanistan to the east and Iraq to the west – Pakistan or Iran?
Iran

7. Which country has the capital of Ulaanbaatar – Uzbekistan or Mongolia?
Mongolia

8. The island of Bali is home to the largest proportion of the Hindu minority in which Southeast Asian Country – Malaysia or Indonesia?
Indonesia

9. The Brahmaputra River empties out into what body of water – The Bay of Bengal or the Arabian Sea?
The Bay of Bengal

10. Kashmir is a disputed region between what two South Asian countries – India/Pakistan or Nepal/Afghanistan?
India/Pakistan

11. Israel borders what body of water to the west – The Persian Gulf or the Mediterranean Sea?
Mediterranean Sea

12. Damascus is the capital of what Southwest Asian country – Jordan or Syria?
Syria

13. Azerbaijan, Georgia, and Armenia are all part of what region – Northern Asia or the Caucasus?
Caucasus

14. East Timor gained independence from what country with the highest population of Muslims in the world – Indonesia or Lebanon?
Indonesia

15. Burma borders what country to the southeast – Vietnam or Thailand?
Thailand

16. How many emirates make up the United Arab Emirates – Seven or Eight?
Seven

17. Dushanbe, Astana, and Tashkent are capitals of three countries that are located in what region – Southwest Asia or Central Asia?
Central Asia

18. India's national currency is what – the Rupee or Rupiah?
Rupee

19. The Yangtze River is the third longest river in the world, and the longest in Asia. What East Asian country does it occupy – North Korea or China?
China

20. The Tigris and Euphrates Rivers are located in what present-day country that used to contain Mesopotamia, one of the first civilizations in the world – Iran or Iraq?
    Iraq

21. Pashto and Dari are the two official languages of which South Asian country – Bhutan or Afghanistan?
    Afghanistan

22. The Khyber Pass links Afghanistan to which bordering country which is the sixth most populous country in the world and contains the populous cities of Karachi and Lahore – Pakistan or Turkmenistan?
    Pakistan

23. Ankara is the capital of which country that had a dispute with Greece over ownership of Cyprus, a now independent nation – Turkey or Akrotiri?
    Turkey

24. SAARC is an Asian Organization that comprises of eight countries from which region – South Asia or the Middle East?
    South Asia

25. Petra, an archaeological city, is located in what Middle Eastern country that is landlocked by Israel, Syria, Saudi Arabia, and Iraq with the capital city of Amman – Jordan or Lebanon?
    Jordan

26. The Republic of China is known by what other name – Hong Kong or Taiwan?
Taiwan

27. Which country is the only doubly landlocked country in the world besides Liechtenstein – Tajikistan or Uzbekistan?
Uzbekistan

28. Ho Chi Minh City is the most populous city in what country – the Philippines or Vietnam?
Vietnam

29. Angkor Wat, a Hindu Temple and the largest religious monument in the world is located in what country whose main religion is Theravada Buddhism – Cambodia or Singapore?
Cambodia

30. Tagalog is a language spoken mainly in which country – Mongolia or Philippines?
Philippines

31. The Gulf of Aden is south of which Middle Eastern country on the Arabian Peninsula – Yemen or Oman?
Yemen

32. Kuwait was invaded by Iraq in what year – 1990 or 1992?
1990

33. Which sea is the Earth's lowest elevation on land – Red Sea or Dead Sea?

Dead Sea

34. Sinhala and Tamil are the two official languages of what country – Maldives or Sri Lanka?
Sri Lanka

35. The 2022 FIFA World Cup will be hosted by what Middle Eastern country – Qatar or Bahrain?
Qatar

36. The Irtysh River is located in what country – Kazakhstan or Kyrgyzstan?
Kazakhstan

37. The mouth of the Mekong River is in what sea?
South China Sea

38. The ancient Phoenicians lived in which country – Lebanon or Israel?
Lebanon

39. Mount Everest is located on the border of China and what other country – India or Nepal?
Nepal

40. The Baikonur Cosmodrome belongs to Russia but is situated in what other country – Kazakhstan or Kyrgyzstan?
Kazakhstan

41. Kyrgyzstan occupies part of what mountain range – Ural or Tian Shan?
Tian Shan

42. Which East Asian country is bordered by the most number of other countries – China or Mongolia?
China

43. Name the plateau situated between the Eastern and Western Ghats – Deccan Plateau or Tibetan Plateau?
Deccan Plateau

44. Which country is an example of an archipelago – Maldives or Sri Lanka?
Maldives

45. Bahrain's national currency is what – Dinar or Apsar?
Dinar

46. Zoroastrianism originated in which country – India or Iran?
Iran

47. Pyongyang is a city in which country – North Korea or South Korea?
North Korea

48. Macau and Hong Kong are Special Administrative Regions of what country where the Yellow and Yangtze Rivers are found – Thailand or China?
China

49. The Mekong River runs through China, Myanmar, Thailand, Cambodia, Vietnam, and what other country – Laos or Malaysia?
Laos

50. The fourth largest religion in Malaysia is what – Hinduism or Jainism?
Hinduism

51. The third largest island in the world is located in Indonesia. What is its name – Sumatra or Borneo?
Borneo

52. The Baha'i Faith's greatest number of followers reside in what South Asian country – India or Pakistan?
India

53. The Himalaya Mountains are formed at the convergent boundary between the Indo-Australian Plate and what other tectonic plate – Arabian Plate or Eurasian Plate?
Eurasian Plate

54. The Petronas Towers in a Southeast Asian country whose currency is the ringgit. Name this country – Malaysia or Singapore?
Malaysia

55. The Indian state of Sikkim borders Nepal to the west and what country to the east where Hinduism and Buddhism both prosper as the main religions – Bhutan or Bangladesh?

Bhutan

56. The Angara River is located in the Irkutsk Oblast in what country – China or Russia?
Russia

57. Yin and Yang are from what religion founded by Lao-Tzu in Ancient China – Taoism or Shintoism?
Taoism

58. The landlocked country of Kyrgyzstan contains what capital – Bishkek or Ulaanbaatar?
Bishkek

59. Syria borders what sea to the west – Black Sea or Mediterranean Sea?
Mediterranean Sea

60. Russian is a language spoken in which of these countries – North Korea or Kazakhstan?
Kazakhstan

61. The Andaman and Nicobar Islands are part of what country – India or Burma?
India

62. West and East Azerbaijan are located in what country that borders Afghanistan and Pakistan to the east – Iran or Iraq?
Iran

63. The Arabian Sea is west of India, and the Bay of Bengal is in what position relative to India – East or South?
East

64. Mohenjo-Daro and Harappa were cities along what river civilization – Ancient India or Ancient China?
Ancient India

65. Taiwan's capital is what – Guangzhou or Taipei?
Taipei

66. The Chao Phraya River flows into the Gulf of Thailand and is a major river of what country – Thailand or Cambodia?
Thailand

67. The Cagayan River and Rio Grande de Mindanao are the largest river systems in what country – East Timor or Philippines?
Philippines

68. Nagorno-Karabakh and what country use the dram as their official currency – Armenia or Georgia?
Armenia

69. Dzonghka is the official language of what South Asian Buddhist country bordering only India and China – Bhutan or Nepal?
Bhutan

70. Borneo, Sumatra, and Java are all very large islands in what Southeast Asian country that is the largest archipelagic country in the world – Malaysia or Indonesia?
Indonesia

71. Turkmenistan borders what country to the northeast – Tajikistan or Uzbekistan?
Uzbekistan

72. The Tibetan Plateau can be found in the country of China, north of what mountain range with mountains that are the highest in the world – Himalayas or the Kunlun Shan?
Himalayas

73. Manila is the capital of what country east of the South China Sea and is separated from Taiwan by the Luzon Strait – Philippines or Vietnam?
Philippines

74. Cambodia is southeast of what country whose capital is Bangkok and contains the Mekong River – Laos or Thailand?
Thailand

75. Buddhism is practiced by a little over 70% of the population of what country – Maldives or Sri Lanka?
Sri Lanka

76. Guangzhou is a city in what East Asian country famous for its large population and size – China or Japan?
China

77. Naypyidaw and Rangoon are cities in what South Asian country bordering India, China, Bangladesh, Thailand, and Laos – Cambodia or Burma?
Burma (Myanmar is acceptable)

78. Adam's Bridge connects two countries, India and what other country – Sri Lanka or Maldives?
Sri Lanka

79. Angora used to be the capital of what country that contains the populous city of Istanbul – Syria or Turkey?
Turkey

80. Balochistan is a province in what country that claimed Bangladesh as part of it until Bangladesh's independence in 1971 – Afghanistan or Pakistan?
Pakistan

81. Inner Mongolia is a state of what country – China or North Korea?
China

82. Mandarin is spoken mainly in what country – Taiwan or China?
China

83. What Chinese city can be found at the confluence of the Jialing and Yangtze Rivers – Chongqing or Guangzhou?
Chongqing

84. The Malay Peninsula is found in what Asian Region – Central Asia or Southeast Asia?
Southeast Asia

85. Papua New Guinea borders what Southeast Asian country – East Timor or Indonesia?
Indonesia

86. Lampung and North Sulawesi are provinces in what country bordering the Indian and Pacific Oceans – Indonesia or Malaysia?
Indonesia

87. The Philippine sea is connected to the South China Sea by what strait – Luzon Strait or Bering Strait?
Luzon Strait

88. Yemen and Oman are countries on what Peninsula – Malay Peninsula or Arabian Peninsula?
Arabian Peninsula

89. The Kamachatka Peninsula is east of what body of water – Sea of Okhotsk or Bering Sea?
Sea of Okhotsk

90. The Amur River forms part of the border between China and what other country – Mongolia or Russia?
Russia

91. The Brahmaputra River empties out into what body of water – Andaman Sea or Bay of Bengal?

Bay of Bengal

92. The Persian Gulf is connected to what sea – Red Sea or Arabian Sea?
Arabian Sea

93. Mongolia is completely surrounded by what two major countries in size – China/Russia or Kazakhstan/Kyrgyzstan?
China/Russia

94. Which country is not landlocked – North Korea or Nepal?
North Korea

95. Which country that is made up of islands spans more degrees of longitude – Seychelles or Maldives?
Maldives

96. Which country is connected to Sri Lanka by a chain of small islands called Adam's Bridge – Malaysia or India?
India

97. Most gerbils sold as pets in the United States are offspring of animals imported from the desert that covers much of Mongolia and part of China. Name this desert – Gobi or Taklimakan?
Gobi

98. A giant statue of Buddha at the confluence of the Min and two other Asian rivers was carved out of a cliff to protect boatmen from dangerous currents. This statue is in what officially atheist country – China or Vietnam?

China

99. Most Palestinians live in a territory called the West Bank and in a country directly to the east. Name this neighboring country – Lebanon or Jordan?
Jordan

100. The Indus River originates in what region of China – Tibet or Yunnan?
Tibet

101. The Ganges River's largest tributary is what river – Brahmaputra or Yamuna River?
Yamuna River

102. The Tigris River merges with the Euphrates to form a river emptying out into what gulf – Gulf of Oman or Persian Gulf?
Persian Gulf

103. Jagged limestone peaks, known as karst formations, are found near the Li River in which Asian country?
China

104. Which region, home to Muslims and Hindus, has been the subject of a territorial dispute for 60 years?
Kashmir

105. In March 1998 heavy rains in the province of Baluchistan caused flooding in which South Asian country that borders the Arabian Sea?

Pakistan

106. The Kara Sea is located north of what country?
Russia

107. Wang is a popular last name from what country that borders the Gulf of Tonkin?
China

108. Lake Baikal borders the Buryat Republic to the southeast, a part of what country with the ethnic groups of Tatar and Bashkir?
Russia

109. Local languages like Gurung and Tamang are spoken in the capital of a country home to Ama Dablam. Name this country.
Nepal

110. Edmund Hillary and what other mountaineer became the first to reach the summit of Mount Everest?
Tenzing Norgay

111. Over 80% of what country with the Murghab and Tejen Rivers is covered by the Karakum Desert?
Turkmenistan

112. The Taiga can be found in Russia and what other East Asian country?
Mongolia

113. The Yangtze River is the longest river in Asia, located entirely within what country?
China

114. The Lena River connects Lake Baikal to what ocean?
Arctic Ocean

115. The Amu Darya is formed by the junction of the Vakhsh and what other river?
Panj River

116. The Salween River empties out into the Andaman Sea and originates in what plateau?
Tibetan Plateau

117. The Irrawaddy River is the longest river in what country located in South Asia?
Burma (Myanmar is acceptable)

118. The Yenisei River begins at the city of Kyzyl in what country?
Russia

119. Luang Prabang, a city at the confluence of the Mekong and Nam Khan Rivers, can be found in what country?
Laos

120. The Krishna and Godavari Rivers are two major rivers in what country?
India

121. The Jordan River in the Middle East flows into what sea?
Dead Sea

122. The Syr Darya originates in the Tian Shan in what Asian region?
Central Asia

123. The Qizilqum Desert in Uzbekistan extends into what country to the north where the Baikonur Cosmodrome is operated?
Kazakhstan

124. The Turpan Depression is located in the northwestern region of a country and is also the country's lowest point. Name this country.
China

125. Name the deepest lake in the world.
Lake Baikal

126. Kanchan Kalan is the lowest point in what South Asian country that is predominantly Hindu?
Nepal

127. Bekaa Valley is located in what Middle Eastern Mediterranean country bordering Syria and Israel?
Lebanon

128. Gangkar Puensum is the highest peak in what country whose primary language is Dzongkha?
Bhutan

129. The Hengduan Shan Mountains are located in China and what other country with the major city of Mandalay and the Shan Plateau?
Burma (Myanmar is acceptable)

130. The Orontes River, stretching into Lebanon and Turkey, can also be found in what country with the cities of Aleppo and Latakia?
Syria

131. Bishkek, the capital of Kyrgyzstan, is located on the edge of what mountain range that extends into China?
Tian Shan

132. What country bordering the Persian Gulf to the east and with the dinar as its official currency is a major producer of oil and is home to Muslims, Christians, and Hindus?
Kuwait

133. Foho Tatamailau is the highest point in what country that gained independence from Indonesia on May 20, 2002?
East Timor

134. Adana is a city in what country bordering Syria to the south?
Turkey

135. What is the only country in Southeast Asia that does not border the sea?
Laos

136. The Cathedral of Notre Dame is located in what country bordering the South China Sea?
Vietnam

137. What country has the highest literacy rate in South Asia?
Sri Lanka

138. From the mid-1600s to the mid-1800s, a city on Kyushu was the only Japanese port open to foreign trade. Name this port city.
Nagasaki

139. The largest number of Azeris live in Azerbaijan and what other country?
Iran

140. Hindus who live in the most populous Muslim country revere Mount Agung and believe it is the center of the universe. This volcano is in what island country?
Indonesia

141. Tigers once lived along the Syr Darya, a river that flowed to an inland sea. What is the name of this sea?
Aral Sea

142. The manat is the currency of what country in the Caucasus whose major exports are oil, gas, machinery, and cotton and was previously part of the Soviet Union?
Azerbaijan

143. What country, whose former name was Dilmun, has the dinar as its official currency and is located east of Saudi Arabia and northwest of Qatar?
Bahrain

144. Keokradong is the highest point in what country containing the Sundarbans and the Ganges River delta?
Bangladesh

145. Rub al Khali is a desert in what country bordering the Gulf of Masira and the Arabian Sea?
Oman

146. Khan Tangiri, located in the Tian Shan, is also the highest peak in what country containing the Ustyurt Plateau?
Kazakhstan

147. Which body of water lies between the Arabian Peninsula and Iran?
Persian Gulf

148. The Khyber Pass is located in what mountain range in southeastern Afghanistan?
Safed Koh Range

149. The ngultrum is the currency of what landlocked country bordering the Indian state of Sikkim to the west?
Bhutan

150. Multan is located in the eastern part of what country whose natural hazards include earthquakes and floods and the tribal area of Azad Kashmir?
Pakistan

151. Aragats, located in the Lesser Caucasus, is the highest point in what country containing Sevan Lake and the Aras River?
Armenia

152. What country in Asia makes up part of a mainland peninsula and part of the island of Borneo?
Malaysia

153. Many Dravidian temples are found in Chennai, a seaside city located in the southern part of which Asian country?
India

154. One of the world's largest hydroelectric dams spans the Angara River at Bratsk. This dam is in what country?
Russia

155. Kuala Lampur, the capital of Malaysia, is at the confluence of the Gombak and what other river?
Klang River

156. Shwedagon Pagoda, a structure topped with 5,448 diamonds, can be found in what Burmese city?
Yangon (Rangoon is acceptable)

157. Archery is the national sport of what landlocked and mountainous South Asian country?
Bhutan

158. Every year approximately 70,000 merchant ships cross between the Strait of Malacca, which is the most direct route between India and China. This strait is situated between Malaysia and what other country?
Indonesia

159. The Makkah Clock Royal Tower is a gigantic structure and the second tallest in Asia. This tower is located in what country on the Arabian Peninsula?
Saudi Arabia

160. The Dead Sea, the lowest point in Asia and the lowest in the world on land, can be found between Israel and what other country bordering Iraq?
Jordan

161. What country, located in Southeast Asia, is the most densely populated country in Asia?
Singapore

162. Irkutsk and Ulan-Ude are Russian cities near what major lake?
Lake Baikal

163. Bangalore is a major city in what country that is the third largest in Asia, after Russia and China?
India

164. What Asian country traditionally organized its citizens into a strict caste system related to Hinduism?
India

165. The Gobi Desert is the main physical feature in the southern half of a country known also as the homeland of Genghis Khan. Name this country.
Mongolia

166. What sea in the Arctic Ocean separates the Taymyr Peninsula from the archipelago Novaya Zemlya?
Kara Sea

167. The Ganges River Delta can be found in what country?
Bangladesh

168. Kaiseki, a multi-course meal with an artistic presentation, is popular in what former capital city located on the Kamo River?
Kyoto

169. Tarbela, a large embankment dam, is located in what country that shares borders with Iran, Afghanistan, China, and India?
Pakistan

170. Dagestan, a region that includes many different ethnic groups, is located just west of what large body of water?
Caspian Sea

171. The Gulf of Oman is to the Arabian Sea as the Gulf of Tonkin is to WHAT?
South China Sea

172. From what northern Indian town, in a province of the same name, can one glimpse Mount Everest?
Darjeeling

173. In May 2008, the largest Chinese earthquake in over 50 years occurred in Sichuan Province. Name the capital of this province, which is located about 50 miles from the epicenter of the earthquake.
Chengdu

174. Name the westernmost national capital in Asia.
Ankara

175. Pakistan's largest province has the country's smallest provincial population. Name this province.
Baluchistan

176. Lake Assad, created by a major hydroelectric dam, is located on which river?
Euphrates River

177. Mawsynram, India, has a lengthy monsoon season and is often referred to as the wettest place on Earth. On average, how many inches of rainfall does Mawsynram receive annually?
467 inches

The Geography Bee Ultimate Preparation Guide

178. Yekaterinburg lies on the eastern side of the Ural Mountains in which country?
Russia

179. Which of these Asian countries has a drier climate – Yemen, the Philippines, or Bangladesh?
Yemen

180. The Ginza and Shinjuku are famous districts in one of the world's most densely populated cities. Name this city.
Tokyo

181. The Maldives are located off the southwest coast of what country on the mainland of Asia?
India

182. Peshawar, a city in the North-West Frontier Province of Pakistan, has had strategic importance for centuries because of its location near what historic pass?
Khyber Pass

183. The 2012 Nuclear Security Summit took place in March in what Asian capital city located on the Han River?
Seoul

184. Hoover Dam is to the Colorado River as Atatürk Dam is to what?
Euphrates River

185. What river, whose drainage basin is the largest in Sri Lanka, and which reaches the Bay of Bengal on the

southwestern side of Trincomalee Bay, is the largest in the country?
Mahaweli River

186. The Fergana Valley is a region in Central Asia spread across Uzbekistan, Tajikistan, and what other country?
Kyrgyzstan

187. The Kara Darya is a tributary of what major river – Amu Darya or Syr Darya?
Syr Darya

188. What country was historically called Formosa by the Portuguese and can be found in the Pacific Ocean?
Taiwan

189. The Kowloon Peninsula forms the southern part of what territory belonging to China?
Hong Kong

190. The Euphrates River begins in Turkey, and flows south, meeting with the Tigris River in what country?
Iraq

191. Paro Airport is the only international airport in what country containing the Paro Valley and the town of Wangdue Phodrang?
Bhutan

192. Cape Dezhnev forms the easternmost mainland point of Eurasia. This cape is on the Chukchi Peninsula in what country?
Russia

193. The Palk Strait separates the Indian state of Tamil Nadu from the Northern Province in what country?
Sri Lanka

194. The Deccan Plateau, located between two major mountain ranges, can be found in what country bordering the Lakshadweep Sea?
India

195. The Jaffna Peninsula, a region inhabited primarily by Tamils, forms the northern tip of what island country?
Sri Lanka

196. Which country does not border India – Nepal, Pakistan, or Japan?
Japan

197. The island of Sulawesi can be found in what country containing the cities of Denpasar and Medan?
Indonesia

198. The Karatal River, in Kazakhstan, is 242 miles long. This river originates in the Dzungarian Alatau Mountains near Kazakhstan's border with China, and flows into what major lake that is the 13th largest continental lake in the world?
Lake Balkhash

199. Dena, is southwestern Iran, is the highest point in what mountain range that spans the whole length of the western and southwestern Iranian Plateau?
Zagros Mountains

200. Lake Tai is a large freshwater lake in the Yangtze Delta Plain. This lake belongs to Jiangsu Province in which country?
China

201. The Pothohar Plateau, in Pakistan, was once home to the ancient Soanian culture. This plateau can be found in what Pakistani province?
Punjab

202. The Jhelum River, which flows through India and Pakistan, is a tributary of what river whose source is the Bara Lacha Pass?
Chenab River

203. The Saigon River, found in Vietnam, rises near Phum Daung in the southeastern part of what country bordering Vietnam?
Cambodia

204. The Gobi Desert, in Mongolia and China, borders what major Asian mountain range to the north?
Altai Mountains

205. Mount Rinjani, in the province of West Nusa Tenggara, is an active volcano on Lombok in what country?
Indonesia

206. An isthmus connects Asia with which other continent?
Africa

207. Name the channel formed by the confluence of the Tigris and Euphrates Rivers that flows into the Persian Gulf.
Shatt al Arab

208. Which country, made up of more than 13,000 islands, has the second largest area of tropical rain forest after Brazil?
Indonesia

209. Phnom Aural is the highest peak in what country situated entirely in the tropical Indomalayan ecozone?
Cambodia

210. Which country is not part of the Arabian Peninsula – Djibouti, Oman, or Qatar?
Djibouti

211. Which country is not located on part of the Indochina Peninsula – Cambodia, Thailand, or Latvia?
Latvia

212. Mt. Damavand, the highest point in Iran, can be found in what mountain range in the northern part of the country?
Elburz Mountains

213. Mount Apo is the highest peak in the second largest island in the Philippines, bordered by the Davao Gulf and Sulu Sea. Name this island.
Mindanao

214. The Dzavhan River is in what country with the cities of Dalandzadgad and Tsagaannuur and contains Uvs Lake as well as a small portion of the Yenisei River?
Mongolia

215. Samarkand, located on the famous Silk Road, is a popular tourist destination in which Central Asian country?
Uzbekistan

216. The Najd Plateau is north of the Rub al Khali desert, which can be found in Yemen, Oman, UAE, Qatar, and what other country?
Saudi Arabia

217. What country, bordering the Arabian and Laccadive Seas, consists of many coral reefs and atolls?
Maldives

218. The Strait of Hormuz, which forms part of the southern coast of Iran, separates it from what other country with Hindu minorities?
Oman

219. Gujranwala is a city in Pakistan close to the border with what other country?
India

220. The Annam Cordillera, extending into Vietnam, covers most of the eastern region of what country containing the Plateau of Xiangkhoang?
Laos

221. Lop Nur, a lake in western China, is at the eastern end of what basin between the Kunlun Shan, Tian Shan, and Altun Shan Mountain Ranges?
Tarim Basin

222. Mosul is a city in the Ninawa governorate and a chief port on the Shatt al Arab in what country famous for its dust storms and sandstorms?
Iraq

223. What capital city in a country bordering the Arabian Peninsula contains most of the population in that country?
Kuwait City

224. Name the capital city of Taiwan.
Taipei

225. The largest country in central Asia stretches from the Caspian Sea to China. Name this country.
Kazakhstan

226. Name the Muslim country that is bordered by India and Myanmar.
Bangladesh

227. What Southeast Asian country is the largest archipelago in the world?
Indonesia

228. New Delhi is the capital of what country bordering Bangladesh, Pakistan, and Nepal?
India

229. Kanchenjunga is the highest peak in what South Asian country where the majority follow Hinduism?
India

230. Which of these cities is farthest east according to their longitude – Beirut, Tehran, or Colombo?
Colombo

231. The Turpan Depression, an oasis near the northern boundary of the Taklimakan Desert, is the lowest point in which country in Asia?
China

232. Chennai, the city in eastern India formerly known as Madras, is a port on which large bay?
Bay of Bengal

233. The West Bank, the Shatt al Arab, and Kashmir are all disputed regions on which continent – Asia or Africa?
Asia

234. Name the mountain range that extends more than 1,000 miles from Kazakhstan's northern border to the Arctic Ocean.
Ural Mountains

235. Which of the following countries has the longest coastline – Iraq, Iran, or Kuwait?
Iran

236. Which river flows from its source on the Tibetan Plateau through Pakistan before draining into the Arabian Sea – Indus River or Ganges River?
Indus River

237. The capital of the country that makes up more than 70 percent of the Arabian Peninsula is one of the most isolated capital cities in the world. Name this city.
Riyadh

238. The Amu Darya is to the Aral Sea as the Volga River is to what?
Caspian Sea

239. Name the river that flows from its source near Lake Baikal across Siberia to the Laptev Sea.
Lena River

240. Name the world's largest saltwater lake, which has traditionally been a major source of sturgeon for caviar?
Caspian Sea

241. Borneo, one of the largest islands in the world, is considered part of which continent?
Asia

242. Most of the world's highest mountains are part of which mountain system – Himalayas or Andes?
Himalayas

243. The Ural Mountains form the western boundary for which continent?
Asia

244. Which continent includes the landlocked countries of Bhutan and Laos?
Asia

245. The Malwa Plateau and the Kathiawar Peninsula are located in a country containing the Bhabar and Terai belts, which lie near the Himalaya Mountains. Name this country.
India

246. Tabos is a city in what Persian Gulf country?
Iran

247. Birqin and Karamay are cities in the northwest region of what country bordering Burma?
China

248. Which correctly describes the location of Yekaterinburg, Russia – on the Trans-Siberian Railway or west of the Aral Sea?

The Geography Bee Ultimate Preparation Guide

On the Trans-Siberian Railway

249. The coastal regions of which place are LEAST likely to have many fjords – Philippines or New Zealand?
Philippines

250. In what country can you find the Anaimalai Hills and the Chinnar Wildlife Sanctuary, and whose largest crop by economic value is milk?
India

251. The island country of Singapore is located at the tip of what peninsula?
Malay Peninsula

252. Which of the following cities has the least average annual rainfall – Sydney, Belize City, or Kabul?
Kabul

253. Which of these countries borders the most countries – Iraq, Saudi Arabia, or Mongolia?
Saudi Arabia

254. The islands of Madura and Bali and part of what country?
Indonesia

255. What Central Asian capital city is located northwest of the densely populated Fergana Valley?
Tashkent

256. Which country has more arable land – Cambodia or Qatar?

Cambodia

257. Mandalay is located near the edge of current tiger habitat. On what river is this city located?
Irrawaddy River

258. A desert, which has a name meaning "Black Sands," covers more than 70 percent of Turkmenistan. Name this desert?
Karakum Desert

259. The Potala was the home of the Dalai Lama until he fled into exile in 1959. The Potala overlooks what Himalayan city?
Lhasa

260. Tyumen is a city in the western part of the Asian portion of what country?
Russia

261. The Echmiadzin Cathedral, a 1,700-year old church, can be found in what country containing the Hazdan River?
Armenia

262. Mtatsminda Mountain is a famous tourist attraction in what country bordering South Ossetia and Abkhazia?
Georgia

263. Maiden's Tower is located in the Walled City in what Caucasian capital city?
Azerbaijan

264. Naxçivan is an exclave of what country bordering the Caspian Sea?
Azerbaijan

265. Atakule is what Turkish city's highest building?
Ankara

266. The Sidon Soap Museum, located in Sidon, can be found in what Eastern Mediterranean country?
Lebanon

267. Dayr az Zawr and Homs are major cities in what country?
Syria

268. Camel caravans stop at Petra when traveling usually between the Red Sea and what other sea?
Dead Sea

269. You can find ancient rock drawings at Timna Park, a lake in a desert in what country?
Israel

270. Adhari Park can be found in what country off the coast of Saudi Arabia?
Bahrain

271. An exclave of Oman lies north of what country bordering the Strait of Hormuz?
United Arab Emirates

272. Asir National Park is located near the coast of the Red Sea in what country?
Saudi Arabia

273. What country is bordered by Iraq and Saudi Arabia and borders the Persian Gulf?
Kuwait

274. Socotra is an island belonging to what country on the Arabian Peninsula?
Yemen

275. The Arabian Oryx Sanctuary is located in what country straddling the Tropic of Cancer?
Oman

276. Mesopotamia Marshland National Park is located a few miles away from what major river?
Euphrates River

277. The Milad Tower, the sixth tallest tower in the world, is located in what country containing Persepolis?
Iran

278. The Baba Wali Shrine is in what city in Afghanistan located near the Arghandab River?
Kandahar

279. The Karakoram Highway stretches 500 miles long from China to what country?
Pakistan

280. What sea, located on the Kazakhstan-Uzbekistan border, is shrinking and disappearing?
Aral Sea

281. Kugitang Nature Reserve contains 438 dinosaur footprints and is located in what country?
Turkmenistan

282. Lake Ysyk is located in what country containing the Naryn and Chuy Rivers?
Kyrgyzstan

283. Garabogaz Bay is an inlet of what sea?
Caspian Sea

284. The cities of Shungay and Oral are located near what country's border with Russia?
Kazakhstan

285. The Zarafshon Mountain Range is primarily located in what country bordering Afghanistan?
Tajikistan

286. Ramoji Film City, where many Bollywood movies are made, can be found in what city in India?
Hyderabad

287. Sri Lanka's Gathering is when hundreds of Asian elephants come to what national park to find water and food?
Minneriya National Park

288. Coral reefs can be found around what group of islands south of the Nine Degree and Eight Degree Channels?
Maldives

289. The Taj Mahal, a blend of Indian and Persian styles, was built in 1648 and is located in Agra in what Indian state with nearly 200 million people?
Uttar Pradesh

290. The Sundarbans, one of the largest mangrove forests in the world, can be found in India and what other country?
Bangladesh

291. Tangail is a city in what country bordering the largest river delta in the world?
Bangladesh

292. The Kuru River can be found in what country containing the famous city of Paro?
Bhutan

293. The Dzungarian Basin borders the city of Urumqi in what country?
China

294. Taichung is a city on what island?
Taiwan

295. The Three Gorges Dam can be found on what major Chinese River?

Yangtze River

296. Lake Biwa, thought to be over 5 million years old, is the largest lake in what country bordering the Bungo Strait?
Japan

297. The Nemegt Basin covers part of the southern region of what country famous for its nomads?
Mongolia

298. Paektu-San is an extinct volcano and the highest mountain in what country containing the Changjin Reservoir?
North Korea

299. Jeju Island is the largest island in what country containing Songni Mountain National Park?
South Korea

300. The Mergui Archipelago consists of islands dotting the western coast of the southeastern region of what country?
Burma (Myanmar is acceptable)

301. Phong Nga Bay borders what country containing the cities of Surat Thani and Songkhla?
Thailand

302. Inle Lake is a beautiful lake in what country bordering China?
Burma (Myanmar is acceptable)

303. The Annamese Cordillera is a mountain range forming the border between Laos and what other country?
Vietnam

304. Kampuchea is the former name of what country bordering Chhak Kampong Saom?
Cambodia

305. Mount Kinabalu is located in what country bordering the Singapore Strait?
Malaysia

306. Tasek Merimbun is the largest lake in what country on the Malay Peninsula?
Brunei

307. Sentosa can be found in what country bordering the Singapore Strait?
Singapore

308. The Natuna Sea forms the eastern coast of what major island in Indonesia?
Sumatra

309. Makassar is on what Indonesian island?
Sulawesi (Celebes is acceptable)

310. Nino Konis Santana National Park is located on the coast of what country bordering the Timor Sea?
East Timor

311. The Chocolate Hills are more than 1,250 identical dirt hills in Bohol. These famous hills are located in what country containing Mount Mayon, sometimes called the world's most symmetrical volcano?
Philippines

312. The Taklimakan Desert borders what mountain range to the south?
Kunlun Shan

313. What river forms part of the border between North Korea and China?
Yalu River

314. The Nine Degree Channel is south of what group of islands belonging to India?
Lakshadweep

315. What oasis and ancient city in the desert in Syria was on a trade route linking the Roman Empire, India, China, and Persia?
Palmyra

316. The Kolyma River is located in the eastern region of what country bordering Shelikhov Gulf?
Russia

317. The ancient Mesopotamian city of Ur is located in what present-day country?
Iraq

318. Sapporo is a city on what Japanese island that is the second largest in the country?
Hokkaido

319. Siem Reap can be found near Angkor and what major Southeast Asian Lake?
Tonle Sap

320. Mymensingh and Jamalpur are cities east of the Ganges River in what country?
Bangladesh

321. Buir Nuur Lake is shared by Russia and what other country?
Mongolia

322. You can ride on a turtle-shaped boat in Suoi Tien Cultural Theme Park in what country bordering the Gulf of Tonkin?
Vietnam

323. Puncak Mandala is a mountain on what island?
New Guinea

324. The Irrawaddy, Mekong, and Salween Rivers can all be found in what country, where the cities of Taunggyi, Pathein, and Sittwe can be found?
Burma (Myanmar is acceptable)

325. The Crocker Mountain Range can be found in the region of Sabah on the Malay Peninsula in what country?
Malaysia

326. Samar and Leyte are two of the major islands of what archipelagic country bordering the Visayan and Samar Seas?
Philippines

327. Lake Balkhash and Lake Tengizy can be found in what country?
Russia

328. The Sea of Galilee, in Israel, is filled by what river?
Jordan River

329. The Palm Islands can be found off the coast of what country containing the cities of Sharjah and Dubai?
United Arab Emirates

330. The Altay Mountains can be found in Mongolia, Russia, China, and what other country?
Kazakhstan

331. Oedo Island can be found in what country?
South Korea

332. Palangkaraya is a city on what Southeast Asian island?
Borneo

333. Przewalski's Horses are the national symbol of what landlocked country?
Mongolia

334. Ahmedabad and Rajkot are cities in what Indian state?
Gujarat

335. Nanjing Lu is the main shopping street in what populous Chinese city?
Shanghai

336. Bandhavgarh National Park is located in what country?
India

337. Novaya Zemlya borders the Barents Sea and what other sea?
Kara Sea

338. The world's largest open-pit gold mine is at Muruntau in what desert?
Qizilqum Desert

339. What space launch facility is the oldest in the world, located in Central Asia?
Baikonur Cosmodrome

340. Darhad Valley is a region in the northern part of what country?
Mongolia

341. The Hongshui River, in China, empties out into what sea?
South China Sea

342. Nagoya is a city on what Japanese island?
Honshu

343. Sanyer is a city in Turkey located to the northeast of what populous Turkish city?
Istanbul

344. The Gulf of Aqaba borders Egypt, Saudi Arabia, Israel, and what other country?
Jordan

345. What desert, known as the Empty Quarter, is the world's largest sand desert and is larger than France?
Rub al Khali

346. The United Arab Emirates borders the Strait of Hormuz to the north, the Persian Gulf to the west, and what small body of water to the east that is an inlet of the Arabian Sea?
Gulf of Oman

347. The Gulf of Aden borders what large Yemeni island?
Socotra

348. You can find many fields or rice in Rangpur, a city in what country famous for its many floods?
Bangladesh

349. What country's rail system transports four billion passengers every year across approximately 30,000 miles of track?
India

350. The most prominent religion and the second most prominent religion in South Asia are Hinduism and Islam. What is the third most prominent?
Buddhism

351. Palawan is an island bordering the Sulu Sea in what country?
Philippines

352. The Anambas Islands and the Natuna Islands are located south of what sea?
South China Sea

353. Bandung is a city on what Indonesian island?
Java

354. The Flores Sea is south of Celebes. Celebes is another name for what Indonesian island?
Sulawesi

355. An exclave of East Timor lies on what island?
Timor

356. Kalimantan province is located in eastern Borneo in what country?
Indonesia

357. Kozhikode, Thrissur, and Kochi are cities in what Indian state whose main language is Malayalam and has an approximate population of about 37 million?
Kerala

358. Masira is an island in the Arabian Sea that belongs to what country on the Arabian Peninsula bordering the United Arab Emirates?
Oman

359. Chilka Lake is located in what Indian state whose capital is Bhubaneswar and borders the Bay of Bengal?
Orissa

360. Sarygamysh is a depression and the lowest point at 39 ft below sea level in what country?
Turkmenistan

361. The Gulf of Khambat and the Gulf of Kutch both form part of the western coast of what country containing the Kathiawar Peninsula?
India

362. What Indian state borders Nepal whose capital and largest city is Patna?
Bihar

363.

364. Khulna is located in the southwestern region of what country bordering the Bay of Bengal?
Bangladesh

365. The Yablonovyy Mountain Range is located in what country?
Russia

366. Jengish Chokusu, also known as Pobedy Peak, is on what country's border with China?
Kyrgyzstan

367. The Central Makran mountain range is located in what country, one of whose tribal areas is Gilgit-Baltistan?
Pakistan

368. Xiangkhoang is a province belonging to what country whose highest point is Pou Bia, located on the Xiangkhoang Plateau?
Laos

369. The densely populated Kanto Plain is located near Tokyo Bay on what island?
Honshu

370. Najd is a desert basin in what country bordering the Red Sea?
Saudi Arabia

371. Faisalabad is a city in what South Asian country?
Pakistan

372. The Gulf of Mannar forms the western coast of what country whose two major religions are Buddhism and Hinduism?
Sri Lanka

373. The Enguri and Rioni Rivers are located in what country?
Georgia

374. Al Wahj is a city in Saudi Arabia on what major Asian body of water?
Red Sea

375. Ataturk Reservoir is located on the edge of the Taurus Mountains in what country?
Turkey

376. The Savu Sea forms the northern coast of what country bordering the Wetar Strait?
East Timor

377. The North and South Malosmadalu Atolls are located in what country whose major exports are fish and clothing?
Maldives

378. Jeju Island is located in what country where the Naktong River can be found?
South Korea

379. The Brahmaputra River has its source in what Chinese Mountain Range?
Gangdise Mountains

380. Kamet is a high peak in the Himalayas, on what country's border with China?
India

The Geography Bee Ultimate Preparation Guide

Europe

1. The city of Limerick is located in what Western European country – Ireland or the United Kingdom?
   Ireland

2. The English Channel is located between the United Kingdom and what other country – Spain or France?
   France

3. The Kremlin is located in what city that is Russia's capital – Moscow or Minsk?
   Moscow

4. Estonia, Latvia, and Lithuania are known as what – Scandinavia or the Baltic States?
   Baltic States

5. Thessaloniki is a city in what Mediterranean country – Greece or Italy?
   Greece

6. The smallest independent country in the world borders Italy. Name this country that holds a population of approximately 800 people – Vatican City or San Marino?

Vatican City

7. What landlocked country is surrounded completely by Austria and Switzerland – Liechtenstein or Luxembourg?
Liechtenstein

8. Crete is an island belonging to what country that was in dispute with Turkey over Cyprus – Greece or Bulgaria?
Greece

9. Krakow, or Cracow, is the second largest city in what country – Lithuania or Poland?
Poland

10. Copenhagen is the capital of what country in Scandinavia – Denmark or Sweden?
Denmark

11. Finland borders what body of water to the south – Gulf of Finland or Gulf of Bothnia?
Gulf of Finland

12. Estonia borders what country to the east – Russia or Latvia?
Russia

13. Longyearbyen is the capital of what Norwegian territory – Jan Mayen or Svalbard?
Svalbard

14. Vienna is the capital of what country where Wolfgang Amadeus Mozart was born – Austria or Switzerland?

Austria

15. The Aland Islands belong to what country – Hungary or Finland?
Finland

16. Andorra is located in what mountain range – Alps or Pyrenees?
Pyrenees

17. Bucharest is to Romania as Budapest is to what – Czech Republic or Hungary?
Hungary

18. Albania borders what country to the east – Macedonia or Croatia?
Macedonia

19. Gibraltar is a territory of what country that is separated from France by the English Channel – United Kingdom or Jersey?
United Kingdom

20. Stonehenge is found in what Western European country – the United Kingdom or Ireland?
United Kingdom

21. Lake Peipus, which empties through the Narva River into the Gulf of Finland, forms part of the border between Russia and which eastern European country – Estonia or Latvia?

Estonia

22. Canary Wharf, part of the Docklands redevelopment, is located on what European river that rises in the Cotswold Hills – Thames or Volga?
Thames River

23. Which correctly describes the location of Aberdeen, Scotland – West of the Grampian Mountains or on the North Sea?
West of the Grampian Mountains

24. Many legal terms come from the language of ancient Rome. Name this language – Latin or Altaic?
Latin

25. Name the westernmost capital city on the mainland of Europe – Lisbon or Madrid?
Lisbon

26. Which of the following is NOT considered one of the Nordic countries – Sweden, Iceland, or Latvia?
Latvia

27. Lederhosen, which are thick leather shorts with suspenders, are part of a costume worn especially in the Bavarian Alps in Austria and which other country – Switzerland or Germany?
Germany

28. An agreement between the Roman Catholic Church and the government of Italy created one of the world's smallest countries. Name this country – Vatican City or Liechtenstein?
Vatican City

29. Which mountains extend across more degrees of latitude – Apennines or Ural Mountains?
Ural Mountains

30. The residents of Andorra, a mountainous country located between Spain and France, have the longest life expectancy in the world. Name the mountain range that dominates Andorra's landscape – Pyrenees or Apennines?
Pyrenees

31. Several countries in central Africa use a unit of currency known as the CFA franc. Most of these countries are former colonies of which European country – France or Belgium?
France

32. Bulgaria borders what country to the north where most of the population is Christian – Romania or Hungary?
Romania

33. The Canary Islands belong to what Western European country that borders France, Andorra, and Portugal – Spain or Monaco?
Spain

34. Rotterdam is the second largest city in what country also known as Holland – Belgium or the Netherlands?
Netherlands

35. Normandy is part of what country that contains the Seine River – Luxembourg or France?
France

36. Sicily, the largest island in the Mediterranean Sea, belongs to what country that borders San Marino – Italy or Vatican City?
Italy

37. Mount Blanc de Courmayeur is the highest point in what country shaped like a boot – Malta or Italy?
Italy

38. Lake Ohrid and Lake Prespa are located in what country north of Greece – Macedonia or Bulgaria?
Macedonia

39. What is the currency of Iceland – krona or euro?
Krona

40. The Crimean Peninsula is part of what country that used to belong to the Soviet Union – Moldova or Ukraine?
Ukraine

41. Cardigan Bay, bordering Wales, is an inlet of what sea – Irish Sea or North Sea?
Irish Sea

42. The Oder Estuary, containing Szczecin Lagoon, is shared by Germany and what other country – Poland or Denmark?
Poland

43. Monaco, the most densely populated country in the world, is bordered by one country. Name this country containing the cities of Toulouse and Marseille and the Seine River – France or Spain?
France

44. Hungarian is closely related to what language spoken in Finland – Swedish or Finnish?
Finnish

45. The Salisbury Plain in southwestern England is the site of what famous Neolithic monument – Big Bentley or Stonehenge?
Stonehenge

46. Galdhopiggen is the highest point in what country on the Scandinavian Peninsula that borders the North Sea, Norwegian Sea, and Barents Sea – Norway or Finland?
Norway

47. The Balearic Sea, which feeds into the Mediterranean Sea, forms much of the eastern coast of what country bordering Andorra and home to the last remaining Iberian Lynx – Portugal or Spain?
Spain

48. Liechtenstein is a doubly-landlocked country bordered by Switzerland to the west and what country to the east that gained its independence in 1156 from Bavaria – Austria or Croatia?
Austria

49. The Gulf of Gdansk forms the northeastern coast of what country that contains part of Stettiner Haff (Szczecin Lagoon) – Poland or Germany?
Poland

50. To see the monument to Queen Victoria at the gates of Buckingham Palace, you would travel to which country – France or the United Kingdom?
United Kingdom

51. Which island is not an overseas department of France – Martinique, Reunion, or Tasmania?
Tasmania

52. Sweden's highest peak, Kebnekaise, is located in the Kjölen Mountains. These mountains run along Sweden's boundary with which country – Norway or Denmark?
Norway

53. Which country is more densely populated – Netherlands or Russia?
Netherlands

54. One of the world's longest vehicular tunnels was built through the highest mountain in the European Alps. Name this mountain – Matterhorn or Mount Blanc?
Mount Blanc

55. In terms of area, Ukraine is the largest country that is located entirely within which continent – Europe or Africa?
Europe

56. The archipelago Islas Malvinas, also called the Falkland Islands, which belongs to the United Kingdom, lies off the southeastern edge of which continent – Asia or South America?
South America

57. Dzyarzhynskaya Hara is the highest point in what Eastern European country previously part of the Soviet Union and containing the Dnieper and Nyoman rivers – Ukraine or Belarus?
Belarus

58. The Meuse River can be found in what country whose languages are Dutch, French, and German, whose currency is the euro, and also contains the major cities of Antwerp and Brussels – Belgium or Netherlands?
Belgium

59. Riu Runer and Coma Pedrosa are the lowest and highest points in what country sandwiched between the populous countries of France and Spain?
Andorra

60. Silesia, one of Europe's principal coal-producing regions, is mostly in what country?
Poland

61. The Dinaric Alps, which extend into Croatia, can be found in what country that has a small border with the Adriatic Sea, was part of former Yugoslavia, and whose currency is the marka?
Bosnia and Herzegovina

62. The Elbe and Morava rivers are located in what country whose currency is the koruna and gained independence in 1993?
Czech Republic

63. The Strait of Otranto forms the southwestern coast of what country?
Albania

64. Constance is a lake in the northwestern part of what country with the cities of Graz, Linz, and Salzburg?
Austria

65. The oracle of Zeus at Dodona is located in which European country?
Greece

66. Dalmatians get their name from Dalmatia, a region along the Adriatic Sea in what country?
Croatia

67. The Brandenburg Gate can be found in the capital city of a country in Western Europe. Name this capital city.
Berlin

68. The largest metropolitan area in Europe can be found in what capital city containing the Kremlin and St. Basil's Cathedral?
Moscow

69. What river, located in Russia, is the longest in Europe and flows through Nizhniy Novgorod?
Volga River

70. Monaco, the most densely populated country in Europe and the world, borders one country containing the Seine and Rhone rivers. Name this country.
France

71. What is the largest country entirely in Europe, with the city of Dnipropetrovs'k?
Ukraine

72. Lake Ladoga, in Russia, is closest to what other country bordering the Gulf of Finland and Baltic Sea?
Finland

73. The Strait of Dover connects the North Sea to what channel?
English Channel

74. What small country, the world's most densely populated, lies on the Côte d'Azur near the France-Italy border?
Monaco

75. Genoa is to the Ligurian Sea as Amsterdam is to WHAT?
The North Sea

76. The Azores islands and the Madeira Islands are territories of which country?
Portugal

77. The Great Dividing Range is to Australia as the Grampian Mountains are to what – Scotland or Germany?
Scotland

78. The Black Sea Lowland, extending into Russia and the Ukraine, can also be found in what landlocked Eastern European country once part of the Soviet Union?
Moldova

79. The Gulf of Taranto forms part of the southeastern coast of what country with the city of Turin and the peak of Dufourspitze?
Italy

80. The Skagerrak Channel separates Norway from what country bordering both the Baltic and North Seas and containing the islands of Sjaelland and Fyn?
Denmark

81. The Strait of Bonifacio forms part of the southern coast of the island of Corsica, which belongs to what country with the cities of Nantes and Nice?
France

82. Musala is the highest point in Bulgaria, in what mountain range?
The Rila Mountains

83. The Great Afold covers most of northern Siberia. It also extends into Romania, Croatia, and what country with the capital city of Budapest?
Hungary

84. The Sea of Hebrides, feeding into the Atlantic Ocean, can be found on the northwestern coast of what country with the cities of Glasgow and Aberdeen?
United Kingdom

85. St. Peter's Square and St. Peter's Basilica are two places you can visit in the least populous country in the world. Name this country surrounded by Rome on all sides.
Vatican City

86. The Apennines cover much of what country that is an exporter of building stone and lime and whose lowest point is Torrente Ausa?
San Marino

87. The Swiss franc is the official currency of what country containing the peak of Grauspitz, on its border with Switzerland?
Liechtenstein

88. Kaunas is a city in what country containing the Venta and Nemunas Rivers?
Lithuania

89. Lolland is an island located south of Sjaelland belonging to what country containing the city of Odense on the island of Fyn?
Denmark

90. Vestfiroir and Sudhurland are regions of what country with the city of Thingvellir and the Blanda River?
Iceland

91. Benelux combines the names of the Netherlands, Belgium, and what landlocked country where languages spoken are German and French?
Luxembourg

92. The Vah River is a tributary of the Danube, and has its source in what major European mountain range?
Carpathian Mounains

93. Bratislava is on the western side of what landlocked country located close to Neusiedler Lake and bordering Hungary to the south?
Slovakia

94. Kirklareli is a city on the European part of what transcontinental country?
Turkey

95. The Tisza River forms part of the boundary between Romania and what other country bordering the Sea of Azov?
Ukraine

96. The Crimean Peninsula borders the Black Sea. This peninsula belongs to what country with the administrative division of Luanska?
Ukraine

97. The Gulf of Cadiz is connected to the Alboran Sea, in the Mediterranean by what strait south of Gibraltar and north of Morocco?
Strait of Gibraltar

98. The Tagus River forms part of the border between Spain and what other country whose languages are Portuguese and Mirandese?
Portugal

99. Christianity is the major religion in what country west of Bulgaria, north of Greece, south of Kosovo and Serbia, and east of Albania?
Macedonia

100. The Gulf of Riga borders Estonia and what other Baltic State whose capital is the chief port on the Daugava River?
Latvia

101. Shannon and Limerick are cities in what country west of the Irish Sea?
Ireland

102. Cephalonia is an island off the western coast of what country with the historic cities of Corinth and Olympia?
Greece

103. Lapland is a region in Norway, Sweden, Russia, and what other country whose main religion is Evangelical Lutheran and whose capital is a chief port on the Baltic Sea and Gulf of Finland?
Finland

104. Lake Peipus can be found between Estonia and what other country containing the Kremlin?
Russia

105. The Czech Republic's highest point is Snezka at 5,256 feet. What river contains the country's lowest point?
Elbe River

106. The Gulf of Bothnia forms the western coast of what country that contains the Aland Islands?
Finland

The Geography Bee Ultimate Preparation Guide

107. The republics of Estonia, Latvia, and Lithuania are collectively known by a name shared with the body of water the countries border. Name this body of water.
Baltic Sea

108. Lake Constance is bordered by Germany, Austria, and what other country?
Switzerland

109. The Rubik's cube was invented in what landlocked country in Central Europe?
Hungary

110. Dresden, a city that has been rebuilt since World War II, is situated on what river?
Elbe River

111. Name the two large islands separated by the Strait of Bonifacio.
Corsica and Sardinia

112. Oysters are harvested near Ostend, a seaport located along the North Sea coast of West Flanders in which country?
Belgium

113. Fountains of ash rained down on the town of Catania when Mount Etna erupted in October 2002. Name the island where Mount Etna is located.
Sicily

114. Tower Bridge is to London as the Bridge of Sighs is to what?

Venice

115. An open-air museum east of Interlaken preserves the French, German, and Italian heritage of which mountainous European country?
Switzerland

116. Which country on the Scandinavian Peninsula is largest in area?
Sweden

117. The instrument—which shepherds once played to soothe their herds—originated in the mountainous regions west of Košice in which central European country?
Slovakia

118. Name the mountains that extend across much of Wales from the Irish Sea to the Bristol Channel.
Cambrian Mountains

119. The Ebro River, which flows into the Balearic Sea, is in which European country?
Spain

120. Tourists often toss coins into the Trevi Fountain at the heart of what city?
Rome

121. Timis County shares its name with a tributary of the Danube and is located in the western part of which European country?

Romania

122. Name two of the three largest sections of Denmark, which include its mainland peninsula and two largest islands.
Jutland, Sjaelland, and Fyn

123. What city on the Elbe River is the capital of the state of Saxony?
Dresden

124. Designed to resemble a ship, the Guggenheim Museum in Bilbao is located near what large bay?
Bay of Biscay

125. The Milan Cathedral lies in the valley of Italy's longest river. Name the river.
Po River

126. St. Petersburg is located on the delta of the Neva River where it empties into what gulf?
Finland

127. The Tyrrhenian Sea separates the island of Sardinia from the mainland of which country?
Italy

128. Parmesan cheese takes its name from Parma, a city in the northern part of what peninsular country?
Italy

129. Crayfish parties, called Kraftskiva, are popular throughout Scandinavia in late summer, including what city located on the easternmost bay of Lake Malaren?
Stockholm

130. The Khatanga Gulf borders what large Russian peninsula south of the Seernaya Zemlya island group?
Taymyr Peninsula

131. The La Mancha and Somontano regions are to Spain as the Champagne and Burgundy regions are to what?
France

132. Lesotho is to the Drakensberg Mountains as Andorra is to the what?
Pyrenees

133. The Erzgebirge Mountains can be found in the northwestern region of what country whose major natural hazard is floods and contains the Vltava River, a tributary of the Elbe River?
Czech Republic

134. Eifel is a mountainous region in what country bordering the Baltic Sea and containing the cities of Bremen and Hannover?
Germany

135. The Black Sea Lowland, extending into Moldova and Ukraine, can be found in the southeastern region of what country containing the Prut River?

Romania

136. Maanselka is a plateau region covering an area from the northeastern to southeastern parts of what country, home to the Oulu River?
Finland

137. What Baltic capital city is a chief port on the Daugava River and is located in the central part of the country?
Riga

138. The Gulf of Gdansk, which feeds into the Baltic Sea, borders the Russian exclave of Kaliningrad and what other country containing the Sudeten Mountains in the southwest?
Poland

139. The North Comino Channel and the South Comino Channel surround Comino, and island belonging to what country that also contains the island of Gozo and Marsaxlokk Bay?
Malta

140. The Gulf of Corinth feeds into what other gulf forming part of the western coast of Greece and feeds into the Ionian Sea?
Gulf of Patrai

141. Italy's chief port is on a gulf of the Ligurian Sea. Name this port city.
Genoa

142. Which European country has a larger population – Serbia or Germany?
Germany

143. Name the seismically active mountain range that runs most of the length of Italy.
Apennines

144. Which country borders the Baltic Sea – Latvia or Romania?
Latvia

145. In August 2002, the city of Prague was hit by a flood that raised the waters of what river that runs through the city?
Vltava River

146. The United Kingdom consists of England, Wales, Northern Island, and what other political unit?
Scotland

147. Holland is a synonym for what country on the North Sea?
Netherlands

148. Which country has more boreal forests – Finland or Costa Rica?
Finland

149. Which country, known for its fjords, has the longest coastline in Europe – Finland or Norway?
Norway

150. The Strait of Dover separates what two countries?

The United Kingdom and France

151. Whitewashed houses are a common site for tourists visiting the Cyclades islands. The Cyclades are part of what European country?
Greece

152. Which of the following is not an island group off the coast of Scotland – Shetland Islands, Orkney Islands, or Falkland Islands?
Falkland Islands

153. The Dardanelles and Bosporus are examples of what kind of physical feature?
Strait

154. In terms of area, Ukraine is the largest country that is located entirely within which continent?
Europe

155. Öland, Corsica, and Malta are islands that are part of which continent?
Europe

156. Which country is not landlocked – Paraguay, Italy, or Central African Republic?
Italy

157. Mont-Saint-Michel, an abbey off the Normandy coast, is in an inlet for the Gulf of Saint-Malo. This gulf is an extension of what larger body of water?

English Channel

158. The Brandenburg Gate, which takes its name from a region that was once part of Prussia, is an architectural landmark in what city?
Germany

159. The city of Gomei is in what country?
Belarus

160. Skogafoss Waterfall is one of the highest waterfalls in what country whose capital is Reykjavik?
Iceland

161. Lolland, Falster, and Zealand are islands in what country bordering Skagerrak and Kattegat?
Denmark

162. Jostedalsbreen Glacier, the largest in Europe, is located in what country containing the cities of Bergen and Stavanger?
Norway

163. Forests cover 70 percent of the land of what country containing thousands of lakes and the major cities of Tampere and Espoo?
Finland

164. Estonia and Latvia border what major gulf?
Gulf of Riga

165. The Irbe Strait separates Latvia from what other country?
Estonia

166. Daugavpils is located in what country containing the Venta River?
Latvia

167. Klaipeda is a city on the western coast of what country bordering the Courland Lagoon and containing the Courland Spit?
Lithuania

168. The Seaside Open-Air Museum is located in Ventspils, which is located close to the House of Crafts. Ventspils is located in what Baltic country?
Lithuania

169. Heilbronn and Cologne are cities in what country bordering Stettiner Haff and containing the North Frisian Islands?
Germany

170. Zugspitze, the highest peak in the Bavarian Alps, is located in what country that is home to the Black Forest?
Germany

171. What country bordes Lake Constance to the west and Neusiedler See to the east?
Austria

172. What is the name of the largest ice caves in the world, located in Austria?

Eisriesenwelt Ice Caves

173. The Vltava and Morava Rivers are located in what country where you can find Castle Lucen?
Czech Republic

174. Slovakia borders what major European mountain range to the north?
Carpathian Mountains

175. The Danube River forms part of the border with Hungary and what other country to the north?
Slovakia

176. Lake Balaton is located in what country containing the cities of Miskolc and Budapest?
Hungary

177. Belarus borders Ukraine, Russia, Latvia, Lithuania, and what other country to the west?
Poland

178. Odesa is a city in what country containing the Kremenchuk and Kakhovka Reservoirs?
Ukraine

179. Which of these countries has the northernmost capital city – Japan, Norway, or Iran?
Norway

180. Elba is an island belonging to what country separated from Malta by the Malta Channel and separated from the French island of Corsica by the Strait of Bonifacio?
Italy

181. Helsingor's strategic location on a narrow strait allowed Danish kings to collect tolls from passing ships. Name the strait.
Oresund

182. Mariupol, a city located at the mouth of the Kalmius River, is located on what sea that is an arm of the Black Sea?
Sea of Azov

183. Arabs from Africa named a large rock Gibraltar as they landed on what large European peninsula?
Iberian Peninsula

184. Basques are a group of people who live primarily in the northern foothills of what country?
Spain

185. What river borders more European countries than any other river on that continent?
Danube River

186. Denmark borders what country?
Germany

187. The Sea of Azov, which borders Ukraine and Russia, is an inlet of what larger inland sea?

Black Sea

188. Andorra is located in what mountain range?
    Pyrenees

189. Medina is to Saudi Arabia as Assisi is to what?
    Italy

190. The Aegean, Ionian, and Adriatic Seas are all part of which larger sea?
    Mediterranean Sea

191. Skrokkur Geyser, which erupts every four to eight minutes, is located in what country?
    Iceland

192. Bornholm is an island belonging to what country containing Mariager Fjord and Randers Fjord?
    Denmark

193. Jostedalsbreen Glacier can be found in what Scandinavian country bordering Skagerrak?
    Norway

194. Foteviken Viking Reserve is located in what country whose population is approximately 9 million and borders the Baltic Sea?
    Sweden

195. Tampere and Espoo are major cities in what country containing thousands of lakes and the Tankar Lighthouse?

Finland

196. Hiiuma and Saaremaa are the two largest islands belonging to what country bordering Russia and containing the cities of Narva and Tartu?
Estonia

197. The Western Dvina River is located in what country bordering Lithuania and the Irbe Strait?
Latvia

198. What country contains part of Courland Spit, borders Courland Lagoon, and is home to the Venta River?
Lithuania

199. Brandenburg is a region is the eastern portion of what country, home to Neuschwanstein Castle and Miniatur Wunderland?
Germany

200. Swarovski Crystal Worlds is located in what country bordering Lake Constance and Neusiedler See?
Austria

201. Ahvenanmaa is a region in what country whose northernmost cities include Utsjoki and Karigasniemi?
Finland

202. The Strait of Messina, which connects the Tyrrhenian Sea with the Ionian Sea, lies between mainland Italy and what large island?

Sicily

203. Friesland is south of the North Sea and Helgolander Bay. This region extends into the Netherlands and can also be found in what country?
Germany

204. The Azores islands and the Madeira Islands are territories of which country?
Portugal

205. The Wicklow Mountains dominate the eastern region of what country whose highest point is Carrauntoohil?
Ireland

206. The Guadiana River has its mouth in the Gulf of Cadiz in Spain and what other country?
Portugal

207. Barriers at Woolwich protect what capital city against storm surges from the North Sea?
London

208. Rocks form symbolic ships that mark Viking burial sites in the country that separates the North Sea and another body of water called the Kattegat. Name this country.
Denmark

209. Ghajn Tuffieha Bay borders what island in the Mediterranean?
Malta

210. The Kama Reservoir is located in what large European country?
Russia

211. L'viv is a city in the western region of what country bordering the Black Sea and Moldova?
Ukraine

212. The Outer Hebrides Islands belong to what country bordering Moray Firth?
United Kingdom

213. Corsica is north of what large Italian island?
Sardinia

214. The Balearic Islands belong to what country where the cities of Valladolid and Vigo can be found?
Spain

215. The French city of Lyon is located on what European river?
Rhone River

216. What sea is south of the Irish Sea, bordering Ireland, Wales, and England?
Celtic Sea

217. Vasterbotten is to Sweden as Osterbotten is to what?
Finland

218. Vatnajokull can be found in what country whose highest point is Hvannadalshnukur at 6,923 feet?
Iceland

219. Lim Fjord and Nissum Fjord can be found on Jutland in what country?
Denmark

220. The Finnsmark Plateau is located in what country whose currency is the krone?
Norway

221. Lake Malaren is located in what country bordering Finland?
Sweden

222. Olavinlinna Castle is a fortress built in the late 1500s. This site is in what country, home to Tankar Lighthouse and the cities of Tampere and Kokkola?
Finland

The Geography Bee Ultimate Preparation Guide

Africa

1. By using the Suez Canal, ships avoid sailing around which continent – Africa or Europe?
   Africa

2. Although Sierra Leone's name has Portuguese origins, what is the country's official language – English or Spanish?
   Spanish

3. Addis Ababa is the capital of what landlocked country bordering Somalia to the east – Ethiopia or Eritrea?
   Ethiopia

4. Togo borders what country to the east – Ghana or Benin?
   Benin

5. What is another name for Cote d' Ivoire – Cape Verde or Ivory Coast?
   Ivory Coast

6. What river, located in Northeast Africa, is the longest in the world – Nile River or Amazon River?
   Nile River

7. Western Sahara is a disputed region in what region of Africa – West Africa or Southern Africa?
   West Africa

8. Kaukau Veld is a lowland region in the northeastern part of what country that has Windhoek as its capital – Namibia or Angola?
   Namibia

9. Sapitwa is the highest point in what country containing part of the Great Rift Valley – Rwanda or Malawi?
   Malawi

10. Jebel ech Chambi is the highest peak in what country bordering the Mediterranean Sea, Algeria, and Libya – Tunisia or Egypt?
    Tunisia

11. Swaziland, a landlocked country in southern Africa, borders South Africa and what other country – Mozambique or Botswana?
    Mozambique

12. The Macina Swamp is located in the central region of what country whose capital is Bamako – Mauritania or Mali?
    Mali

13. The Serengeti Plain, extending into Tanzania, is also located in what country to Tanzania's north that borders Lake Victoria – Kenya or Uganda?
    Kenya

14. What country in West Africa, formerly belonging to France, contains the capital of Conakry and has the franc as its official currency – Sierra Leone or Guinea?
Guinea

15. Islam, Hinduism, and what other religion with nearly 2.3 billion followers make up most of Zambia's population – Christianity or Buddhism?
Christianity

16. Four African countries—Chad, Niger, Nigeria, and Cameroon—border which lake located south of the Sahara – Lake Victoria or Lake Chad?
Lake Chad

17. The Gulf of Guinea can be found west of what country that occupies the island of Bioko and contains Pico de Santa Isabel – Equatorial Guinea or Cameroon?
Equatorial Guinea

18. The Senegal River forms all of the border between Senegal and what other country whose lowest point is Sebkha de Ndrhamcha – Mali or Mauritania?
Mauritania

19. Melekeok is the capital of what country whose highest point is Mount Ngerchelchuus?
Palau

20. Hauts Plateaux, which extends into Morocco, is also found in the northern part of what country – Libya or Algeria?
Algeria

21. The Namib Desert, which extends into Namibia, can also be found in the southwestern part of what country bordering the Atlantic Ocean – Angola or Gabon?
Angola

22. Porto-Novo is the capital of what country in western Africa bordering Togo to the west and Nigeria to the east – Ghana or Benin?
Benin

23. The Ahaggar and Atlas mountains can both be found in what country containing the Great Erg Occidental – Morocco or Algeria?
Algeria

24. Lake Eyasi can be found in Tanzania in what major African valley – Great Rift Valley or Kerio Valley?
Great Rift Valley

25. Río Muni is the mainland portion of the only African country that has Spanish as its official language. Name this country – Guinea or Equatorial Guinea?
Equatoral Guinea

26. The Panama Canal is to the Caribbean Sea and Pacific Ocean as WHAT is to the Red and Mediterranean Seas – Erie Canal or Suez Canal?

Suez Canal

27. Maseru is the capital of what landlocked country in southern Africa – Swaziland or Lesotho?
Lesotho

28. What country, bordering only Senegal, contains the Gambia River and the city of Banjul – Gambia or Liberia?
Gambia

29. Which island is larger in Sao Tome and Prinicpe – Sao Tome or Principe?
Sao Tome

30. Melilla and Ceuta are cities claimed by Spain in what country – Algeria or Morocco?
Morocco

31. Mount Tahat, the highest peak in Algeria, can be found in what major mountain range – Atlas Mountains or Ahaggar Mountains?
Ahaggar Mountains

32. Tunisia borders what sea to the north – Mediterranean Sea or Alboran Sea?
Mediterranean Sea

33. The port cities of Al Bayda and Misratah can be found in what country bordering Egypt to the west – Libya or Chad?
Libya

34. Laayoune is a major city in what disputed Moroccan territory – Western Sahara or the Madeira Islands?
Western Sahara

35. Nouakchott is a port on the Atlantic Ocean. It is also a city in what country south of Western Sahara and west of Mali – Senegal or Mauritania?
Mauritania

36. Air Massif can be found in what country with the cities of Niamey and Agadez?
Niger

37. The Gambia River flows through Gambia, Guinea, and what other country bordering Mali and Mauritania?
Senegal

38. What country is located entirely within Senegal except for a small part bordering the Atlantic Ocean?
Gambia

39. Conakry is a capital city located on the coast of what country?
Guinea

40. What country is bordered by Guinea and Liberia and borders Grain Coast?
Sierra Leone

41. Buchanan is a city in what country with the capital city of Monrovia?

Liberia

42. Abidjan is the administrative capital of Cote d'Ivoire, or Ivory Coast. What is the legislative capital, located within the mainland of the country?
Yamoussoukro

43. Ouagadougou is the capital of what landlocked country north of Ghana located in Sahel?
Burkina Faso

44. Accra, on the Gulf of Guinea, is the capital of what country containing Lake Volta and the major city of Kumasi?
Ghana

45. Benin borders Nigeria to the west and what similarly shaped country to its east bordering the Gulf of Guinea?
Togo

46. What city in Benin is the seat of government, on the Gulf of Guinea?
Cotonou

47. Abuja, Ibadan, Kaduna, Kano, Aba, and Maiduguri are major cities in the most populous country in Africa, home to approximately 180 million people. Name this country, bordering Cameroon to the west.
Nigeria

48. Djenne, the oldest known city in Sub-Saharan Africa, can be found in what landlocked country?

Mali

49. What gulf east of Egypt borders Israel to the north?
Gulf of Aqaba

50. Omdurman is a city west of Khartoum North, a city in what country bordering the Red Sea and containing the Nubian Desert?
Sudan

51. Massawa is on the coast of what country that gained independence from Ethiopia in 1993?
Eritrea

52. Laka Tana and Blue Nile Falls are major landmarks in what country that contains the second largest population in Africa and one of the top 20 in the world?
Ethiopia

53. Djibouti borders the Red Sea to the north and what Gulf to the east that has the same name as a port city in Yemen?
Gulf of Aden

54. Hargeisa is the capital of what disputed region currently in Somalia and borers the Gulf of Aden?
Somaliland

55. Uganda borders Lake Albert, Lake Edward, and what major lake to the southwest that is shared with Kenya and Tanzania?
Lake Victoria

56. Rwanda borders what lake to the northwest that lies east of Democratic Republic of the Congo?
Lake Kivu

57. Lake Rudolf, in northern Kenya, is more commonly known as what?
Lake Turkana

58. Mount Kenya is the second highest peak in Africa, located in what country with the port city of Mombasa?
Kenya

59. Dodoma is the legislative capital of what country bordering Lake Tanganyika and Lake Malawi?
Tanzania

60. Bujumbura, the capital of Burundi, is close to what lake?
Lake Tanganyika

61. The Aswan High Dam can be found in what country with Lake Nasser in the southern region?
Egypt

62. Sudan claims boundaries in Egypt and what other country?
Kenya

63. The Tibesti Mountains can be found in the northern part of what country containing Lake Chad and the Chari River?
Chad

64. Douala is a port city on what gulf west of Cameroon, Equatorial Guinea, and Gabon?
Gulf of Guinea

65. Rio Muni can be found in what country with the cities of Bata and Malabo?
Equatorial Guinea

66. Franceville and Lambarene are cities found in what country bordering the Angolan exclave of Cabinda to the south?
Gabon

67. Bangui, the capital of the Central African Republic, lies closest to what neighboring country?
Democratic Republic of the Congo

68. Kinshasa lies close to what other capital city on the other side of the Congo River?
Brazzaville

69. Boyoma Falls, known as Stanley Falls, can be found in the northwestern region of what country containing the Congo and Lomami Rivers?
Democratic Republic of the Congo

70. Annobon, an island south of the Gulf of Guinea, belongs to what small country?
Equatorial Guinea

71. Ituri Forest can be found in what country contining the Kasai and Uele Rivers and the city of Lubumbashi?

Democratic Republic of the Congo

72. Lobito, Luanda, and Benguela are port cities on the Atlantic Ocean in what large country?
Angola

73. Livingstone lies very close to the border of Zimbabwe. This city belongs to what landlocked country with the cities of Kabwe and Chingola?
Zambia

74. Lilongwe is the capital of what country bordering Lake Malawi?
Malawi

75. The Zambezi River is connected to the Indian Ocean through what country?
Mozambique

76. Maromokotro is a peak in what country in the Indian Ocean separated from Mozambique by the Mozambique Channel?
Madagascar

77. Mayotte, located southeast of Comoros and northwest of Madagascar, is a territory of what Western European country?
France

78. Moroni, the capital of Comoros, lies closest to what country?
Mozambique

79. What country in Africa has the highest percentage of Hindus?
Mauritius

80. Lake Kariba is located in southern Zambia and the northwestern region of what country with the cities of Bulawayo and Chitungwiza?
Zimbabwe

81. The Namib Desert constitutes the entire western part of what country containing the Caprivi Strip and Etosha Pan?
Namibia

82. Makgadikgadi Pans is located in what landlocked country famous for its massive diamond production?
Botswana

83. Kruger National Park, larger than Israel, is located in what country with the cities of Port Elizabeth and Durban?
South Africa

84. The Drakensberg and Maloti Mountain Ranges can be found in South Africa and what other country whose lowest elevation is at 1,400 meters?
Lesotho

85. Mbabane and what other city are the administrative and legislative/royal capitals of Swaziland?
Lobamba

86. Brazzaville stands directly across from what other capital city?
Kinshasa

87. Ancient rock carvings can be found in Saguia el Hamra, a region located in what disputed territory south of Morocco?
Western Sahara

88. Which of these countries is the smallest – Angola, Nigeria, or Kenya?
Kenya

89. Lake Victoria, Lake Malawi, and Lake Tanganyika are found in what major African valley?
Great Rift Valley

90. You can visit downtown Johannesburg, a major city in what country?
South Africa

91. What country, in the Indian Ocean, is the most densely populated in Africa?
Mauritius

92. What city in Western Africa is the continent's most populous metropolitan area?
Lagos

93. Lake Assal, the lowest point in Africa, can be found in what country bordering the Red Sea?

Djibouti

94. What lake, in the Great Rift Valley, borders Kenya, Uganda, and Tanzania and is the largest in Africa?
Lake Victoria

95. Rock paintings in the Tibesti Mountains suggest the region once had a much wetter climate. This region is occupied by nomadic herders in what desert – Sahara or Kalahari?
Sahara

96. Nick Nichols photographed lions near the Seronera River in what national park located west of Lake Natron?
Serengeti National Park

97. The confluence of the Blue Nile and the White Nile Rivers lies near Khartoum in which African country?
Sudan

98. The Temple of Queen Hatshepsut at Deir el-Bahri is on the west bank of which African river?
Nile

99. Malagasy is one of the official languages of which country located in the Indian Ocean?
Madagascar

100. Name the landlocked country that borders Botswana and South Africa.
Zimbabwe

101. One of the oldest universities in the world, al-Azhar, was founded in the tenth century and still exists today in what present-day capital city located near the Eastern Desert?
Cairo

102. Dating from the eighth century, one of the oldest libraries in the Muslim world was located at Zaytuna Mosque in what present-day capital city near the ancient city of Carthage?
Tunis

103. Oyala, a planned city located in the rainforest 65 miles east of Bata, is being built as a capital for which African country?
Equatorial Guinea

104. Aswan High Dam is closest to what capital city?
Cairo

105. The Kikuyu, who led the Mau Mau uprising against the British, are the largest ethnic group in which country in East Africa?
Kenya

106. Name the only African nation that has Spanish as its official language.
Equatorial Guinea

107. Name the peninsula in the Middle East that separates the Gulf of Suez from the Gulf of Aqaba.
Sinai Peninsula

108. What city includes a mosque built on top of the ruins of an ancient temple, with hundreds of sphinxes stretching into the distance?
Luxor

109. Name the easternmost national capital on the mainland of Africa.
Mogadishu

110. The Gulf of Aqaba and the Gulf of Suez both border what peninsula?
Sinai Peninsula

111. Mozambique has a 1,600-mile-long coastline on which ocean?
Indian Ocean

112. The Limpopo River, whose source is in the Drakensberg Mountains, forms the entire border between Zimbabwe and what other country?
South Africa

113. The Awash River, located in the Great Rift Valley, has its mouth in Lake Abbe, in what country whose highest point is Ras Dejen?
Ethiopia

114. Malabo is situated on what island belonging to Equatorial Guinea that contains the cities or Riaba and Luba?
Bioko

115. Mount Karisimbi is in the Virunga Mountain Range, on the border between the Democratic Republic of the Congo and what other country whose lowest point is the Ruzizi River?
Rwanda

116. Benin, a city in the Edo state, is in what country containing the Kainji Reservoir and the Sokoto Plains?
Nigeria

117. What country's largest island borders the Gulf of Guinea to the north and whose population is approximately 185,000?
Sao Tome and Principe

118. Thabana-Ntlenyana is the highest point in what country whose lowest point is the Orange/Makhaleng River Junction?
Lesotho

119. Gabon, Chad, Congo, and the Central African Republic were once part of which region – French West Africa or French Equatorial Africa?
French Equatorial Africa

120. Name the river that provides irrigation for most of the cotton grown in Africa.
Nile River

121. Which continent includes the Atlas Mountains and the Kalahari Desert – Africa or South America?
Africa

122. A swampy region in northern Africa, known as the Sudd, is drained by tributaries of which river – Nile or Zambezi?
Nile

123. Madagascar is the largest island in what ocean?
Indian Ocean

124. Which continent includes the Atlas Mountains and the Kalahari Desert?
Africa

125. Which continent contains the largest number of landlocked countries?
Africa

126. Which of these islands is NOT divided between two or more countries – Madagascar, Papua New Guinea, or Borneo?
Madagascar

127. Which country does not experience a tropical wet climate – Kenya, Indonesia, or Brazil?
Kenya

128. Nigeria has extensive oil reserves in the Gulf of Guinea and in the delta of what major river?
Niger River

129. More than 100,000 people fled to the neighboring countries of Liberia and Guinea during the nine months that a West African country was ruled by a military. Name this country.

Mali

130. Berbers and what other Muslim people make up most of the population in Algeria, Libya, and Morocco?
Arabs

131. If built, the proposed Grand Inga Dam would be the world's largest hydroelectric dam. Near the Inga Falls, it is on which African river?
Congo River

132. What river that flows through Africa is the longest river in the world?
Nile

133. Name the sea that was created by the spreading of the earth's crust along the junction of the African and Arabian plates.
Red Sea

134. The Equator passes through the Congo Basin on what continent?
Africa

135. What country is bordered by Lesotho and Botswana?
South Africa

136. The river that forms part of Mauritania's southern border shares its name with what neighboring country?
Senegal

137. Adrar des Iforas is a plateau region, located in the Sahara in Algeria and what other country?
Mali

138. The Cubango Plateau, which has its source in the Bie Plateau, extends into Namibia from what country?
Angola

139. Maromokotro is the highest peak in what country containing the Sofia and Mahavavy Rivers?
Madagascar

140. Banjul is located at the mouth of what river in West Africa?
Gambia River

141. Hhohho is a district in what country whose cities include Mankayane and Nhlangano, and whose lowest point is the Great Usutu River?
Swaziland

142. Known for its leather goods and carpets, Marrakech is in the central part of which African country?
Morocco

The Geography Bee Ultimate Preparation Guide

Australia and Oceania

1. About one-third of New Zealand's population lives in the area around its largest city. Name this city – Auckland or Wellington?
   Auckland

2. The Atherton Tableland, at the northern end of the Great Dividing Range, is known for its relatively high amount of rainfall. This tableland is located in which Australian state – Tasmania, Queensland, or Victoria?
   Queensland

3. Suva is the capital of what country in Melanesia – Fiji or Vanuatu?
   Fiji

4. Mount Cook is located in the Southern Alps of what country containing Milford Sound – Papua New Guinea or New Zealand?
   New Zealand

5. Wake Island is a coral atoll belonging to what North American country – The United States or Canada?
   The United States

The Geography Bee Ultimate Preparation Guide

6. The Waitomo Caves can be found in what country containing the city of Queenstown, a major tourist attraction – Solomon Islands or New Zealand?
   New Zealand

7. Nuku'alofa is the capital of what country in the South Pacific – Tonga or Tokelau?
   Tonga

8. Adamstown is the capital of what territory of the United Kingdom home to only 56 people?
   Pitcairn Islands

9. The Marshall Islands and Kiribati are countries in what region – Micronesia or Melanesia?
   Micronesia

10. What is the currency of Palau – East Caribbean Dollar or U.S. Dollar?
    U.S. Dollar

11. Funafuti and Vaitupu are atolls belonging to what island country – Tuvalu or Nauru?
    Tuvalu

12. New Caledonia's climate is what – tropical or arid?
    Tropical

13. Kimberly Plateau and Lake Eyre Basin are located in the northwestern and central regions of what country

bordering the Gulf of Carpentaria and containing the administrative division of New South Wales?
Australia

14. Honiara is the capital of an island country in the South Pacific. Name this country.
Solomon Islands

15. Islands in which ocean make up the region known as Oceania?
Pacific Ocean

16. The highest mountain in Papua New Guinea has the same name as the last kaiser, or emperor, of Germany. Name this peak.
Mount Wilhelm

17. Anna Creek Cattle Station, which is slightly larger than Israel, can be found in what Australian state?
New South Wales

18. The Great Barrier Reef can be found off the eastern coast of what country?
Australia

19. What country, with the capital city of Yaren, is the most densely populated in Australia and Oceania?
Nauru

20. Caboolture is located in the Moreton Bay Region just north of which Australian state capital city?

Brisbane

21. In February 1998 a major power failure temporarily crippled the economy of New Zealand's most populous city. Name this city.
Auckland

22. Sakhalin Island is to Russia as New Britain Island is to what?
Papua New Guinea

23. Name the Australian island territory in the Indian Ocean that was named for the day it was seen in 1643.
Christmas Island

24. Dolohmwar is the highest point in what country whose languages include Trukese and Pohnpeian and includes the Caroline Islands?
Federated States of Micronesia

25. Most of central Australia has which type of climate – Mediterranean or arid?
Arid

26. The Murray-Darling River can be found in what country?
Australia

27. After Antarctica, which continent has the fewest people per square mile?
Australia

28. Joseph Bonaparte Gulf borders the Northern Territory and what Australian state?
Western Australia

29. Cape York, at the tip of Queensland, is in what country?
Australia

30. Pegasus Bay and Canterbury Bight border what island in New Zealand?
New Zealand

31. The Huon Peninsula is located in what country?
Papua New Guinea

32. Large ranches occupy a region called the Great Artesian Basin in what country that borders the Indian Ocean?
Australia

33. Arnhem Land, which includes a tract of Aboriginal-owned land in northern Australia, lies west of what gulf?
Gulf of Carpentaria

34. Raikura is the third largest island in a country in Oceania. This island belongs to what country?
New Zealand

35. The Ratak and Ralik Chains are groups of atolls and coral islands in what country?
Marshall Islands

36. The Line Islands are an island group in the eastern part of what country that has maritime borders with Tuvalu and Tokelau?
    Kiribati

The Geography Bee Ultimate Preparation Guide

Antarctica

1. To fly the shortest route from Santiago, Chile, to Perth, Australia, you would fly over which continent – Antarctica or Asia?
   Antarctica

2. Antarctica is located in what circle – the Arctic Circle or the Antarctic Circle?
   Antarctic Circle

3. What station in Antarctica had the coldest temperature on record on Earth – Amundsen-Scott South Pole Station or Vostok Station?
   Vostok Station

4. Excluding Antarctica, which continent is the least populous – South America or Australia?
   Australia

5. Antarctica's "population" mainly consists of who – Researchers or Meteorologists?
   Researchers

6. What continent lies closest to Antarctica – Australia and Oceania or South America?
South America

7. Vinson Massif, the highest point in Antarctica, is located in what mountain range – Transantarctic Mountains or Ellsworth Mountains?
Ellsworth Mountains

8. Mount Erebus is an active volcano on what island – Deception Island or Ross Island?
Ross Island

9. Plateau Station in Antarctica belongs to what country – United States or Russia?
United States

10. The Antarctic Treaty was signed in 1959 by how many countries – thirteen or twelve?
Twelve

11. What explorer named Mount Erebus after one of his ships – Roald Amundsen or James Clark Ross?
James Clark Ross

12. What ocean surrounds Antarctica completely – Southern Ocean or Indian Ocean?
Southern Ocean

13. What is it called at the place where the Pacific, Atlantic, and Indian Oceans meet the Antarctic Circumpolar Current – The Antarctic Convergence or the Antarctic Peninsula?
    The Antarctic Convergence

14. Cape Crozier borders the Ross Sea and what other body of water – McMurdo Sound or Prydz Bay?
    McMurdo Sound

15. Bentley Subglacial Trench is the lowest point in Antarctica, bordering Marie Byrd Land and what mountain range – the Shackleton Mountains or Ellsworth Mountains?
    Ellsworth Mountains

16. Roosevelt Island is located on what ice shelf – Ross Ice Shelf or Ronne Ice Shelf?
    Ross Ice Shelf

17. The Riiser-Larsen Peninsula borders what bay to the southeast – Porpoise Bay or Lutzow-Holm Bay?
    Lutzow-Holm Bay

18. Joinville Island can be found off the coast of what peninsula – Antarctic Peninsula or Riiser-Larsen Peninsula?
    Antarctic Peninsula

19. Mt. Minto can be found on what cape – Cape Adare or Cape Poinsett?
    Cape Adare

20. Alexander Island borders what sea – Weddell Sea or Bellingshausen Sea?
Bellingshausen Sea

21. What percentage of Antarctica is not covered by ice?
2%

22. In Dry Valleys, Antarctica, what blasts away snow – gravity or high winds?
High winds

23. Geothermal pools, attracting about 15,000 tourists every year, can be found on what island?
Deception Island

24. The Antarctic pearlwort grows along what major peninsula close to South America?
Antarctic Peninsula

25. The highest recorded temperature in Antarctica was at 59 degrees Fahrenheit in what year?
1974

26. The highest point in Antarctica is Vinson Massif, and the lowest point in Antarctica is what trench?
Bentley Subglacial Trench

27. What is the name of the coldest place in Antarctica, with an annual average temperature of -94 degrees Fahrenheit?
Ridge A

28. What is the number of year-round research stations in Antarctica?
    39

29. Graham Land can be found on what peninsula in Antarctica?
    Antarctic Peninsula

30. Roosevelt Island is surrounded by an Antarctic ice sheet that shares its name with what nearby sea – Weddell Sea or Ross Sea?
    Ross Sea

31. The Bellingshausen Sea and the Ross Sea border which continent – Australia or Antarctica?
    Antarctica

32. The South Shetland Islands, which are claimed by the United Kingdom, Chile, and Argentina, are off the coast of which continent – Antarctica or Asia?
    Antarctica

33. The Larsen Ice Shelf is located on what major peninsula?
    Antarctic Peninsula

34. Vinson Massif is located in what Antarctic mountain range?
    Ellsworth Mountains

35. Carney Island and Siple Island are very close to what ice shelf about 200 miles away from Marie Byrd Land?
    Getz Ice Shelf

36. The Riiser-Larsen Peninsula juts out into what ocean other than the Southern Ocean?
Indian Ocean

37. The Ronne Ice Shelf borders what major Antarctic Sea?
Weddell Sea

38. Bransfield Strait borders the Palmer Archipelago and Graham Land. Graham Land is on what peninsula?
Antarctic Peninsula

39. Victoria Land borders the Transantarctic Mountains, Prince Albert Mountains, and what minor mountain range bordering Cape Adare?
Admiralty Mountains

40. The West Ice Shelf and Shackleton Ice Shelf both border what sea?
Davis Sea

41. Mount Erebus is a volcano on what continent?
Antarctica

The Geography Bee Ultimate Preparation Guide

Physical Geography

1. What term describes the air circulation pattern of rising air at the Equator and sinking air around 30 degrees latitude – Hadley Cell or westerly?
   Hadley Cell

2. Which commonly serves as the baseline for measuring elevations of Earth's landforms – sea level or river level?
   Sea level

3. What is the term for the process by which ice sheets expand and reshape the physical landscape – Weathering or Glaciation?
   Glaciation

4. What resource is trapped between layers of rock in an aquifer – Water or Oil?
   Water

5. What is created when lightning suddenly heats the air, causing it to expand rapidly – Thunder or Rain?
   Thunder

6. How many degrees of latitude are there south of the Equator – 90 degrees or 180 degrees?
   90 degrees

7. Seismology is the study of what – Volcanoes or Earthquakes?
   Earthquakes

8. What is the term for a navigable waterway connecting two larger bodies of water – Strait or Bay?
   Strait

9. What is the term for the lower limit of permanent snow cover on a mountain – snowline or treeline?
   Snowline

10. Which is the term for the frozen ground in Arctic areas where soil temperatures remain below freezing most or all of the year – permafrost or desert?
    Permafrost

11. A meteorologist is to weather as a geologist is to WHAT – rocks or oceans?
    Rocks

12. The tropopause marks the boundary between the troposphere and which other layer of the atmosphere – stratosphere or mesosphere?
    Stratosphere

13. What is the term for a huge desert area characterized by deep sand dunes – erg or harmattan?
Erg

14. What is the term for a narrow projection of a larger territory?
Panhandle

15. What word describes a plain without trees characteristic of the arctic and subarctic regions?
Tundra

16. The place at which two streams or rivers flow together to form one larger stream or river is known as what?
Confluence

17. What term is used for an isolated hill/mountain with steep sides, having a smaller summit area than a mesa?
Butte

18. What is the term for a crater formed when the roof of a cavern collapses?
Sinkhole

19. What is the term applied to magma when it has appeared on the Earth's surface?
Lava

20. A time of widespread glaciation is called a what?
Ice Age

21. What type of rock forms when molten materials harden?
    Igneous Rock

22. What is the term for a long and narrow inlet with steep cliffs or sides that is created by glacial erosion?
    Fjord

23. What zone in the interior of the Earth is between the crust and the outer core?
    Mantle

24. What term can describe when rocks and soil are carried and deposited by a glacier?
    Moraine

25. What is the term for a line of bold cliffs?
    Palisades

26. An area of lessened or decreased precipitation on the lee side of a mountain range or mountain is known by what term?
    Rainshadow

27. A piece of subcontinental land surrounded on all sides by water is called by what name?
    Island

28. What is the term for a large body of water that lies within a curving coastline, smaller than a sea and usually larger than a bay?
    Gulf

29. What is name of the point where salt water from the ocean meets with freshwater from a river?
Estuary

30. A break in the Earth's crust that can occur along plate boundaries is known as a what?
Fault

31. A long cliff or steep slope separating two level or more gently sloping surfaces, resulting from faulting or erosion is known by what name?
Escarpment

32. What is the term for the study of the surface waters of Earth?
Hydrography

33. What term describes a sandy/rocky surface material deposited by meltwater that flowed from a glacier?
Outwash

34. What is the term for a cone-shaped feature on the floor of a cave that is formed by slow dripping water?
Stalagmite

35. What is the term for the large tropical storms that sometimes take place in the Caribbean Sea?
Hurricanes

36. Which characteristic is shared by all deserts – low precipitation or many sand dunes?
Low precipitation

37. The mangrove tree is to coastlines as the cactus is to what?
Desert

38. The distinctive shape of Europe's Matterhorn was created by which agent of erosion?
Ice

39. The Great Circle, at zero degrees latitude, is known by what name?
Equator

40. What is the term for the sudden vibrations caused by the movement of rock along a fault?
Earthquake

41. Name the belt of volcanic and seismic activity that borders most of the Pacific Rim.
Ring of Fire

42. What is the name of the solid fossil fuel found in sedimentary rock?
Coal

43. In the Northern Hemisphere, what season begins when the noonday sun is directly overhead at the Tropic of Capricorn?
Winter

44. What resource is trapped between layers of rock in an aquifer?
Water

45. Earthquakes can sometimes create unusually large waves that cause destruction when they reach land. What Japanese term is used for this kind of wave?
Tsunami

46. In Florida and the West Indies, a small low-lying island usually made up of coral or sand is known by what term?
Key

47. What is the term for the flat area that stretches beyond the banks of a river?
Floodplain

48. What is the term for a hot spring through which jets of heated water and steam erupt?
Geyser

49. Which of these physical features is most different from the others – reef, marsh, or wetland?
Reef

50. Oceans are to oceanography as fossils are to WHAT – paleography or paleontology?
Paleontology

51. What term describes the drainage pattern on a cinder cone volcano such as Mount Etna?

Radial

52. Agulhas and Kuroshio are both names of what kind of physical feature?
Ocean Current

53. Irrigated fields can lose their productivity when water evaporates, leaving salts that gradually accumulate in the soil. What is the term for this process?
Salinization

54. A compass is to direction as a global positioning system is to what?
Location

55. Partially decomposed organic soil material is known by what name?
Humus

56. What term describes a permanently frozen layer of soil?
Permafrost

57. What is the term for an isolated hill or mountain of resistant rock that rises above an eroded lowland?
Monadnock

58. A circular depression containing a volcanic vent is known by what term?
Crater

59. The innermost layer of Earth is known by what *specific* name?
Inner Core

60. Molten rock that shoots out of a volcano during an eruption is known by what name?
Lava

61. What type of stone is composed mainly of calcite and is a sedimentary rock – Sandstone or Limestone?
Limestone

62. A large mass of rocks/sediment carried and deposited by a glacier typically as ridges at its edges is known by what term?
Moraine

63. A landform created at the mouth of a river, where it separates into distributaries, is also formed from the deposition of sediment carried by the river. Name this landform.
Delta

64. A line on a map connecting points that receive equal precipitation is known by what name – Isohyet or Isobar?
Isohyet

65. A steep-sided volcano built by tephra deposits and lava flows is known by what name besides "composite volcano"?
Stratovolcano

66. Something that can be dissolved is known by what other term?
Soluble

67. A drowned river valley remaining open to the sea is known by what geographical term?
Ria

68. Evaporation and plant transpiration together form what long geographical term?
Evapotranspiration

69. Ice sheets covered part of what is now the United States during the Pleistocene epoch, which took place during Earth's current geologic era. Name the current geologic era.
Cenozoic

70. Earth's diameter at the Equator is slightly larger than its diameter at the Poles. To the nearest whole number, give the equatorial diameter of the Earth in miles.
7926 miles

71. The Arctic Circle is to 66 1/2° north latitude as the Tropic of Cancer is to what?
23 1/2° north latitude

72. What is the term for the fan-shaped feature composed of sand and gravel that is formed where a stream emerges from a mountain valley onto a plain?

Alluvial fan

73. What term is used for the two halves of Earth that are divided by the Equator?
Hemisphere

74. What is the term used to describe a tropical grassland containing widely spaced trees – deltas or savanna?
Savanna

75. What condition is necessary for the formation of coral reefs – warm ocean temperature or annual rainfall of 300 inches?
Warm ocean temperature

76. The prime meridian is which line of longitude – 0 degrees or 90 degrees?
0 degrees

77. Which of the following is not a name for a rotating storm – typhoon, monsoon, or hurricane?
Monsoon

78. Eye is to hurricane as epicenter is to what?
Hurricane

79. What does a barometer measure – wind direction or atmospheric pressure?
Atmospheric pressure

80. Name the belt of volcanic and seismic activity that borders most of the Pacific Rim.
Ring of Fire

81. Cenotes, or sinkholes such as those found on the Yucatán Peninsula, are associated with which rock—basalt or limestone?
Limestone

82. What is the term for the ecosystem generally consisting of broadleaf evergreen trees that is found in wet tropical areas such as the Amazon basin?
Rainforest

83. What agent of erosion is primarily responsible for creating limestone caves?
Water

84. A group or chain of islands is known by what name?
Archipelago

85. A human-made waterway used by ships or to carry water for irrigation is known by what term?
Canal

86. What term describes the side away from or sheltered from the wind?
Leeward

87. A long, narrow piece of land often made of sand or silt extending into a body of water from the land is known as what?
Spit

88. What term can be used for a large, floating mass of ice?
Iceberg

89. A narrow sea inlet enclosed by high cliffs is known by what term?
Fjord

90. A long depression usually created by a river or glacier bordered by higher land is known as a what?
Valley

91. An area of land lower than the surrounding countryside that is usually flat without hills and mountains called a what?
Lowland

92. A point of land extending into an ocean, lake, or river is known by what term?
Cape

93. What is the term describing an area with little or no human settlement?
Bush

94. An elevated area drained by different river systems flowing in different directions is known as a what?

Divide

95. A mound or ridge of wind-blown sand is known by what term?
    Dune

96. What is the term for a lengthy period of time during which thick glaciers cover much of the Earth?
    Ice Age

97. Which of the following do scientists NOT use to study climate change – fault lines or plant fossils?
    Fault Lines

98. Long, narrow fields that resemble stair steps on steep hillsides are commonly known by what name?
    Terraces

99. The sun is directly overhead at noon at the Equator two times each year. What is the term for these events?
    Equinox

100. What term refers to the day in the Northern Hemisphere when the length of time between sunrise and sunset is the shortest of the year and the sun is farthest south of the Equator?
    Winter Solstice

101. In Florida and the West Indies, a small low-lying island usually made up of coral or sand is known by what term?
    Key

102. A plant community dominated by thickets of shrubs and small trees and usually found in Mediterranean climates is known as what?
Chaparral

103. What term do geographers use to express the number of people per square mile of a country's area?
Population Density

104. A mixture of mud and straw is a popular building material in the southwestern United States. What is this building material called in this region?
Adobe

105. What is the name of the solid fossil fuel found in sedimentary rock?
Coal

The Geography Bee Ultimate Preparation Guide

Cultural Geography

1. The Ganges River in South Asia is sacred to followers of what religion that has over 1 billion followers?
   Hinduism

2. Bandar Seri Begawan is the capital of what Southeast Asian country whose official languages are Malay and English – East Timor or Brunei?
   Brunei

3. Kaliningrad Oblast is an exclave of what country that hosted the Sochi 2014 Winter Olympics – Ukraine or Russia?
   Russia

4. The Asian Games is an Asian multi-sport event held every four years in different locations. Which country holds the record for the most gold medals – Japan or China?
   China

5. Nepal and Bhutan are countries both whose two most practiced religions are Hinduism and Buddhism, religions that originated in what present-day country that contains the states of Goa, Bihar, and Madhya Pradesh – India or Bangladesh?

India

6. Mecca and Medina are holy Islamic cities in what country on the Arabian Peninsula – Saudi Arabia or Yemen?
Saudi Arabia

7. The Olympic Games were invented in what country that was home to the ancient Minoans – Greece or Turkey?
Greece

8. The kora, which is a musical instrument made from a gourd, is often played in Gambia on which continent – Africa or Asia?
Africa

9. Muslims in the city of Mopti pray at mosques constructed of clay around a wooden frame. Mopti is southeast of Mauritania in what country – Mali or Senegal?
Mali

10. The principal Christian denomination in Egypt shares its name with a form of the ancient Egyptian language. Name this denomination – Coptic Christian or Roman Catholic?
Coptic Christian

11. Couscous is a popular dish from the northern part of which continent – Africa or Europe?
Africa

12. The Han are the most populous ethnic group in which mainland Asian country – Thailand or China?

China

13. The island of Honshu can be found in what country, home to many followers of Shintoism – South Korea or Japan?
Japan

14. The rumba, a popular ballroom dance in America, was introduced to the United States from which island country – Bahamas, Jamaica, or Cuba?
Cuba

15. Finnish and Swedish are languages spoken in the Aland Islands, a territory of what Nordic country – Finland or Sweden?
Finland

16. The Maori are the indigenous people of what country in Australia/Oceania?
New Zealand

17. Amharic is the official language of what landlocked country in Northeast Africa?
Ethiopia

18. Carnival is an annual festival in what country whose predominant language is Portuguese?
Brazil

19. Jainism is followed mainly in what country in South Asia, home to the Red Fort?
India

20. About 90 percent of the people in which South American country speak Guaraní, an indigenous language – Paraguay or Uruguay?
Paraguay

21. The Huang He is commonly known by what English name?
Yellow River

22. St. Basil's Cathedral, located in Red Square, is an Orthodox church in which country – Ukraine or Russia?
Russia

23. Shintoism is what type of religion – polytheistic or monotheistic?
Polytheistic

24. What is the largest sect of Christianity – Protestant or Roman Catholic?
Roman Catholic

25. The rulers of the Mughal Empire in India were followers of what religion – Hinduism or Islam?
Islam

26. Xhosa and Zulu are languages spoken in what country in Africa that is famous for its penguins – South Africa or Madagascar?
South Africa

27. The Khmer Empire followed mostly Hinduism and a little of what other religion – Buddhism or Confucianism?
Buddhism

28. Kannada is a language spoken in South India, from what language group – Austroasiatic or Dravidian?
Dravidian

29. Mandarin Chinese, Spanish, English, Arabic, Russian, and what other language are the six official languages of the United Nations – French or German?
French

30. Which state has a larger Asian-American population – California or Utah?
California

31. Which language is derived from ancient Celtic – Basque or Welsh?
Welsh

32. Mandarin is to Chinese as Castilian is to WHAT – French or Spanish?
Spanish

33. The origin of Europe's name is often attributed to Europa, a princess in stories from where – Greek Mythology or Roman Mythology?
Greek Mythology

34. Carnival is a festival celebrated in Rio de Janeiro in what country?
Brazil

35. The indigenous people of Australia include the Aboriginal peoples and a second group whose name comes from what strait that separates Australia from New Guinea?
Torres Strait

36. Makossa is a type of music popular in which country that lies northeast of the island of Bioko?
Cameroon

37. What is the official language of Brazil?
Portuguese

38. What is the primary religion of Mali, a landlocked country located in West Africa?
Islam

39. The Festival of San Fermin at Pamplona is an important festival in which European country?
Spain

40. Impressionist painting was developed by painters such as Monet and Renoir in which European country?
France

41. What religion originated in India and spread across Asia via trade routes such as the Silk Road?
Buddhism

42. What country is famous for its harmonious gardens that provide sanctuary for people living in large cities like Kyoto?
Japan

43. The oracle of Zeus at Dodona is located in which European country?
Greece

44. The Manchu are an ethnic group of what Asian country?
China

45. Which of the following countries does not have one of the Romance languages as its official language – Italy, Albania, or Spain?
Albania

46. Kuchipudi is a type of dramatic dance displaying rhythmic footwork and graceful body movements. This classical dance takes its name from a village near the mouth of the Krishna River in what Indian state that borders the Bay of Bengal?
Andhra Pradesh

47. Name the ethnically distinct region of northern France that was settled by Celtic people.
Brittany

48. The Harmandir Sahib, also known as the Golden Temple, is a cultural and sacred center for Sikhs in what state of India?
Punjab

49. Which country does not have French as an official language – Gabon, Haiti, or China?
China

50. Paella, a traditional rice dish, originated near the Albufera Lagoon, just south of what city at the mouth of the Turia River?
Valencia

51. Couscous, a traditional Berber dish made from wheat, is sometimes served with seafood in what African capital city located near Cape Bon?
Tunis

52. Cinco de Mayo is a holiday celebrated in the United States by people who emigrated from which country?
Mexico

53. The Borobudur Temple Compounds, home to hundreds of Buddha statues, is an ancient center for pilgrimage and education in Mahayana Buddhism. This monument, the world's largest Buddhist site, can be found in what country?
Indonesia

54. The incredible complex of Chichen Itza revealed much about the Maya and Toltec's visions of the universe. What famous temple, known as El Castillo, demonstrates the importance of Mayan astronomy?
Temple of Kukulkan

The Geography Bee Ultimate Preparation Guide

55. The face of the sun god adorns a wall of a temple at Kohunlich, an ancient Mesoamerican city in the state of Quintana Roo. This temple is on what peninsula?
Yucatan Peninsula

56. In which Asian country would you find kabuki actors using music and colorful, elaborate costumes to tell a story?
Japan

57. Name the continent on which the Olympic Games originated.
Europe

58. Music of the sitar, a long-necked stringed instrument, is most commonly associated with which country in Asia?
India

59. A large portion of the population in which country does not speak Spanish – Spain, Mexico, or Finland?
Finland

60. Which country's population is not predominantly Muslim – Kuwait, Canada, or Iran?
Canada

61. Theravada Buddhism is the largest religion in what country whose main languages are Khmer, French, and English?
Cambodia

62. Bastille Day is a national holiday celebrated in which European country?

France

63. The rupee is to India as the what is to China?
    Yuan

64. Hundreds of wooden stave churches, containing both Christian and Viking symbols, were built during the Middle Ages in the Nordic country that borders the Barents Sea. Name this country.
    Norway

65. The word "trek," meaning a trip or journey, came into common English usage after the migration of Dutch-speaking people in southern Africa. Name this group of people?
    Afrikaans

66. Native Americans in the southwestern United States impersonate spirits by wearing masks and making dolls. What are these spirits called?
    Kachinas

67. Brahma, Shiva, and Vishnu are three main deities of India's principal religion. Name this religion.
    Hinduism

68. The Star of David is a symbol of Judaism, a religion which has its largest population of followers on which continent?
    North America

69. People in the most populous Scandinavian country celebrate a festival of light called St. Lucia Day to mark the start of the Christmas season. Name this country.
Sweden

70. The lively Juhannus Festival is celebrated in what Nordic country bordering the Gulf of Bothnia and containing the city of Oulo?
Finland

71. Minarets are towers that play an important role in what religion?
Islam

72. The art of making porcelain was first perfected by the people of what ancient civilization?
Ancient China

73. Intricately carved and painted puppets, used in an ancient Southeast Asian form of shadow play, may be seen in religious ceremonies on an island in the Lesser Sundas that is predominantly Hindu. Name this island, where Denpasar is the provincial capital?
Bali

74. Navratri is a major Hindu festival celebrating the different avatars of the goddess Durga for nine days. This holiday is mostly celebrated in what country where Hindi, Tamil, Malayalam, Kannada, Telugu, Marathi, and Urdu are spoken?
India

75. About 90 percent of Poland's people adhere to what religion?
Christianity

76. Which country celebrates an almost 200 kilometer long ice skating race called the Eleven Cities Tour – Netherlands or Denmark?
Netherlands

77. An ancient rock carving shows people skiing some 6,000 years ago. This rock art was discovered in which country that borders the North Sea – Norway or Latvia?
Norway

78. Robot jockeys race camels in Saudi Arabia. It's weird, but it's true! These races take place on which continent that borders the Persian Gulf?
Asia

79. Transylvania is a region that was made famous by a fictional character, Count Dracula. This region is located in which European country?
Romania

80. Which country has four official languages, including Romansh, which is based on ancient Latin?
Switzerland

81. The Forbidden City is a former imperial palace located in the center of which Chinese city – Beijing or Tianjin?

Beijing

82. The Faisal Mosque, the largest mosque in South Asia, is located in which city that is adjacent to Rawalpindi?
Islamabad

83. Although it can trace its influences from many areas of the world, jazz music was created on which continent?
North America

84. The Masai of Kenya and Tanzania base their wealth on which animal – cow or elephant?
Cow

85. What is the chief language of Austria?
German

86. English is the official language in what South American country whose population is roughly 50 percent East Indian?
Guyana

87. Raiatea, one of the Society Islands, was an important religious and cultural center for ancient Polynesians. Today Raiatea is part of what overseas territory?
New Zealand

88. Which Scandinavian country has an official language related to Magyar, Hungary's national tongue?
Finland

89. French patois is spoken in what country whose cities include Soufriere and La Plaine, and which borders Grand Bay?
Dominica

The Geography Bee Ultimate Preparation Guide

Economic Geography

1. Santos, a major coffee-exporting port, is on the Atlantic coast of what country – Costa Rica or Brazil?
   Brazil

2. What is the term for a marketplace in Middle Eastern and North African cities where a variety of goods is sold – Bazaar or Salam?
   Bazaar

3. The Southern Common Market, or Mercosur, was established in 1991 to eliminate trade barriers among several countries on which continent – Africa or South America?
   South America

4. In 2005, severe drought caused a decrease in agricultural production in which country whose capital is Madrid – Greece or Spain?
   Spain

5. Peanuts are the principal export of the African country that is almost completely surrounded by Senegal. Name this country – Gambia, Liberia, or Mauritania?

Gambia

6. Arizona, New Mexico, and Utah are the leading mining states of a metal used in the plumbing and electrical industries. Name this metal – Iron or Copper?
Copper

7. Which South American country has approximately 4,600 miles (7,400 kilometers) of coastline along the Atlantic Ocean, contributing to the large tourism trade in that country – Brazil or Uruguay?
Brazil

8. The production of maple syrup is an important economic activity in which state – Vermont or Utah?
Vermont

9. Coffee is a major export of what country in North America – Haiti or Canada?
Haiti

10. What country is the number one exporter of bananas – Ecuador or Chile?
Ecuador

11. The number one country in uranium production is located in Asia. Name this country.
Kazakhstan

12. What country located in the Middle East and on the Arabian Peninsula is the world's top exporter of oil?

Saudi Arabia

13. The top producer of lentils is what country located in South Asia?
India

14. Nearly 85 million tourists visited a country in 2013 containing the city of Marseille. Name this country.
France

15. What country is the second largest producer of apples?
United States

16. What country, bordering the Indian and Pacific Oceans, is the second largest producer of gold?
Australia

17. What country bordering three oceans contains 536 billionaires, the highest number in the world?
The United States

18. Three of the top five countries leading in bauxite production are located in Asia. The other two are Australia and what other country leading in third?
Brazil

19. What country is the world's top producer of silicon?
China

20. What country is the world's second largest producer of steel?

Japan

21. Sugar production and tourism help give a country on Hispaniola a higher gross national product. Name this country.
Dominican Republic

22. High birthrates and falling death rates would cause which demographic trend?
Population growth

23. Antwerp, because of its location near the North Sea, has become the chief port and commercial center of which country in Europe – Belgium or Denmark?
Belgium

24. The economy of Brunei, a small country on the island of Borneo, is based on offshore oil and natural gas fields that lie beneath the waters of which ocean – Indian or Pacific?
Pacific

25. The Ginza, a shopping and entertainment district, got its name because shoguns used to mint their coins in the area. Name the Asian capital city where the Ginza is located – Beijing or Tokyo?
Tokyo

26. What country is Africa's largest producer and exporter of oil and contains Port Harcourt, the center of the country's oil industry?
Nigeria

27. The Kano Region produces half of what country's peanut crop?
Nigeria

28. The population of Australia rapidly increased in the 1850s as a result of which economic activity – gold mining or sheep ranching?
Gold mining

29. Mining is one of the fastest-growing sectors of the economy of the largest country in the Western Hemisphere. Name this country.
Canada

30. Name the ice-free, deepwater port that is one of Canada's most important ports for trade with Asia.
Vancouver

31. Denim is a fabric named for the city of Nimes, where it was first made. This city is west of the Rhône River in what country?
France

32. In 2001, El Salvador adopted the currency of which country, its largest trading partner?
United States

33. Cotton, fruit, and grain are among the crops grown at El Faiyum, an oasis just south of El Giza, in which country?
Egypt

34. Pine and eucalyptus trees have been planted to support the wood pulp industry in the country that is almost surrounded by South Africa. Name this country.
Swaziland

35. Which of the following was not a chief industry in Great Britain during the 19th century – textile manufacturing, oil refining, or iron and steel production?
Oil refining

36. Which country in South Asia is the world's leading producer of tea?
India

37. What country, made up solely of islands, can be found in the Pacific Ocean and is one of the world's largest exporters of bananas?
Philippines

38. What country, the largest and most populous country in South America, is an airline industry leader?
Brazil

39. About two-thirds of the world's emeralds come from what South American country bordering both the Pacific and Atlantic Oceans?
Colombia

40. In Papua New Guinea, sunken ships from what famous war are popular attractions for scuba divers?

World War II

41. Which country located southeast of Australia is one of the world's leading producers of wool?
New Zealand

42. Oil pipelines cross the Isthmus of Tehuantepec to the port at Salina Cruz in which country?
Mexico

43. Coconut milk is a common drink on an island about 100 miles east of St. Vincent in the Lesser Antilles. Name this island.
Barbados

44. Tropical flowers are sold to tourists who visit the largest island in the Windward island chain of the West Indies. Name this French island.
Martinique

45. The Cu Chi Tunnels, an extensive network of war-time tunnels, are tourist attractions just northwest of a city formerly known as Saigon. Name this city.
Ho Chi Minh City

46. One of the Prairie Provinces produces more oil than all the other Canadian provinces combined. Name this province.
Alberta

47. The fertile Ayeyarwady delta yields a large rice crop in what Southeast Asian country?

Burma (Myanmar is acceptable)

48. What city that lies on the Volga River delta is an important shipping center for the Caspian Sea region?
Astrakhan

49. The discovery of a major shale deposit in the Vaca Muerta formation in 2010 has led to an expansion of oil drilling in the Neuquen province in what country?
Argentina

50. Large reserves of phosphates are found in which North African country that borders the Atlantic Ocean and the Mediterranean Sea?
Morocco

51. Place these countries in order according to their annual oil production, from most to least: Saudi Arabia, Mexico, Canada.
Saudi Arabia, Canada, Mexico

52. Alkmaar, a town famous for its cheese market, is on a canal with access to the Waddenzee. Alkmaar is in which country?
Netherlands

53. Seventy percent unemployment and a severe fuel shortage have virtually paralyzed economic activity in which country east of Botswana?
Zimbabwe

54. One of the world's largest potash deposits is located in Saskatchewan in which country?
   Canada

55. Khark Island is an important oil export terminal in the Persian Gulf. This island belongs to what country?
   Iran

56. Nickel is mined near Pico Duarte, one of the highest peaks in the West Indies. This peak is located in which country west of Puerto Rico?
   Dominican Republic

57. Geothermal springs are an attraction near Rotorua in what country?
   New Zealand

58. Oil refining is an important economic activity on the largest island in the Netherlands Antilles. Name this island.
   Curacao

59. What is the term for a tax that makes goods imported by a country more costly than similar goods produced within that country?
   Tariff

60. Chile has extensive copper reserves in what desert?
   Atacama Desert

61. Uranium mines near the shores of Lake Athabasca are found in two Prairie Provinces in which country?

Canada

62. Chicle, an ingredient in chewing gum, is a sticky substance obtained from a tree native to which country?
Mexico

63. Which country has a higher life-expectancy rate – Australia or Ukraine?
Australia

64. Which country does not have a high population density – Australia or Bangladesh?
Australia

65. Which of these countries has the highest literacy rate – Japan, the Philippines, or Indonesia?
Japan

66. Brisbane and Adelaide are ports on which continent?
Australia

67. A booming economy has given which European country the nickname the "Celtic Tiger"?
Ireland

68. Canada's Prairie Provinces were settled by a large number of immigrants from which country that is also a major wheat-producing region in Europe?
Ukraine

69. What country, whose main exports are oil and gold, contains the island of Bougainville and borders the Coral Sea?
Papua New Guinea

70. Boats called gondolas are used as a means of transportation in which European city – Geneva or Venice?
Venice

71. What domesticated root crop became the staple food of both the Inca Empire and Ireland?
Potato

72. On average, each resident of which industrialized country produces more garbage than the people of any other country – United States or Japan?
United States

73. Sailors on sailing ships, fearful of being becalmed, try to avoid a region where trade winds converge in a low-pressure belt. What is the term for this region?
Doldrums

74. The world's northernmost national capital is heated almost entirely by geothermal energy. Name this city.
Reykjavik

75. Which river delta has been the site of many rebel attacks on pipelines and facilities near rich oil deposits – Niger or Congo?
Niger

The Geography Bee Ultimate Preparation Guide

76. Oil spills from ships traveling through the Bosporus are an environmental concern in what country?
Turkey

77. Which state produces more apples – Washington or Mississippi?
Washington

78. The world's highest-quality emeralds are mined near Bogota in what South American country?
Colombia

79. One of the world's largest producers of nutmeg includes a clove of the spice on its national flag. Name this Caribbean island country.
Grenada

80. Which country is home to Bollywood, the world's largest movie-making industry?
India

81. Which country has a greater amount of proven oil reserves – United Arab Emirates or South Korea?
United Arab Emirates

82. Which beverage crop provides most of the export incomes for Rwanda and Burundi – coffee or tea?
Tea

83. Port Harcourt, a major oil exporting center, is located on the southern coast of which West African country?
Nigeria

84. Which capital city, located on the Mediterranean Sea, is an important port for Libya – Tripoli or Mogadishu?
Tripoli

85. Name the sea that is an important shipping route between Australia and New Zealand.
Tasmanian Sea

86. One of the world's largest known natural gas reservoirs, the North Field, belongs to what country in the Middle East?
Kuwait

87. What waterway serves as a shortcut for most ships traveling from New York to Los Angeles?
Panama Canal

88. What city, once the capital of imperial Russia, is the easternmost port on the Gulf of Finland?
St. Petersburg

89. Other than tourism, what is the primary industry of the Bahamas – Textile or Banking?

90. What European country is dependent on a system of bridges and ferries for transportation between its main peninsula, Jutland, and its Baltic Sea islands?
Denmark

91. The production of maple syrup is an important economic activity in which state – Vermont or Utah?
Vermont

92. Name the city that became Côte d'Ivoire's chief port after it was linked to the Gulf of Guinea by the construction of a canal.
Abidjan

93. The fertile Plain of Sharon is a major citrus-growing region along the Mediterranean coast in what country?
Israel

94. Barley is an important crop grown in East Anglia, the easternmost region in what country?
United Kingdom

95. Lumbering and paper milling are two important economic activities in Missoula, a city located just south of the Flathead Indian Reservation, in which U.S. state?
Montana

96. The production and crushing of copra, the dried meat of coconuts, are important activities in a country that lies on the eastern half of the world's second-largest island. Name this country.
Papua New Guinea

97. What is the term for a city, such as Paris, that has a much larger population than the country's second most populous

city and is the center of economic, political, and cultural activities?
Primate city

98. Bulawayo is a city in what country whose major exports include gold and ferroalloys, and whose highest point is Inyangani at 8,504 feet?
Zimbabwe

99. Banana plantations surround the port city of Puerto Limón, located on the Caribbean Sea in which country?
Costa Rica

100. Traditionally, sugarcane and tobacco have been the chief exports of what country in the Greater Antilles?
Cuba

101. Rio de Janeiro and what other large city in Brazil are growing toward each other due to the high rate of urbanization?
Sao Paulo

102. The Philippines produces more of which grain – wheat or rice?
Rice

The Geography Bee Ultimate Preparation Guide

Historical Geography

1. The most famous library of ancient times was located in what Egyptian city – Cairo or Alexandria?
   Alexandria

2. Mahatma Gandhi fought for India's independence from what country through nonviolence – Australia or the United Kingdom?
   India

3. The *Mayflower* and the *Titanic* are two famous ships that began their transatlantic journeys at what British port city – Southampton or Plymouth?
   Plymouth

4. The Lewis and Clark expedition of 1804 to 1806 began at the mouth of the Missouri River and ended at the mouth of what other North American river – Columbia River or Snake River?
   Columbia River

5. When Christopher Columbus started to sail to India, he landed in another continent, which he thought was India.

This continent was actually populated by what major indigenous people?
Native Americans

6. Althingi, established in the year 930, is the world's oldest continuous parliament in what country that was ruled under the Danish crown for 500 years before becoming independent – Iceland or Greenland?
Iceland

7. European starlings were brought to the U.S. in the 1890s by a man who wanted to introduce every bird mentioned in Shakespeare's plays. He released more than 60 starlings into which city's famous Central Park – Washington D.C. or New York City?
New York City

8. Teotihuacán was the great center of an early civilization on which continent – North America or South America?
North America

9. Lucknow, located on the Gomati River, was the site of a mutiny against the British in 1857 in what present-day South Asian country?
India

10. Which European explorer was the first to see the Pacific Ocean – Balboa or Magellan?
Balboa

11. In what year did Bangladesh gain independence from Pakistan – 1971 or 1981?
1971

12. Vasco de Gama sailed around what continent to get to India – Europe or Africa?
Africa

13. Montezuma was killed by whom – Spanish conquistadors or the British Empire?
Spanish conquistadors

14. The Inca lived on the western coast of what continent – North America or South America?
South America

15. Chichen Itza was a famous monument and is now a tourist attraction in what country – Mexico or Guatemala?
Mexico

16. Rivalry over Korea and what region in China led to the Russo-Japanese War in the early 20th century – Manchuria or Xinjiang?
Manchuria

17. Lewis and Clark traveled through which present-day state as they explored the Louisiana Purchase – North Dakota or Louisiana?
North Dakota

18. In 1768 the Gurkhas captured a city in a Himalayan valley and made it their capital. Name this capital city.
Kathmandu

19. What historical era resulted in the greatest growth of cities in Europe and North America?
Industrial Revolution

20. The Lascaux Caves are a series of cave paintings in what country?
France

21. Which country was ruled by the Tokugawa Shogunate and had a warrior class of Samurai?
Japan

22. Valparaíso, which was founded by the Spanish in the mid-1500s, is the oldest and largest port on the Pacific coast of which country?
Chile

23. Before capturing Samarkand, Alexander the Great crossed the Oxus River, which flowed from the Pamirs to the Aral Sea. Today, the river is known by what other name – Brahmaputra or Amu Darya?
Amu Darya

24. In April 1998 the Shroud of Turin went on display at the cathedral in the city of Turin. This Italian city is located on what major river – Po River or Arno River?
Po River

25. The Northern Territory's largest town south of the Macdonnell Ranges was founded as a station on the Overland Telegraph Line in the late 1800s. Name this Australian town.
Alice Springs

26. The Pillars of Hercules, named after a hero of ancient Greek mythology, mark the eastern entrance of what strategically important strait?
Strait of Gibraltar

27. After the American Revolution, many Loyalists from the southeastern states emigrated with their slaves to Exuma and other islands that today are part of what former British colony?
Bahamas

28. What California city was reduced to ruins following a strong earthquake and disastrous fire in 1906?
San Francisco

29. Which state was the first to secede from the United States in the months leading up to the Civil War – South Carolina or Maryland?
South Carolina

30. Genoa, the city believed to be the birthplace of Columbus, is in which present-day country?
Italy

The Geography Bee Ultimate Preparation Guide

31. The Louvre first opened as a museum in the French Revolution in what year – 1794 or 1793?
    1794

32. A city on the Yangtze River northwest of Shanghai was China's capital when the Communists took control of the country in 1949. Name this city.
    Nanjing

33. In the 1840s the Sultan of Oman relocated his capital to a small island off the coast of East Africa. Name this island, which became a center of ivory and slave trade?
    Zanzibar

34. What town, located at the junction of the Potomac and Shenandoah Rivers, was the site of an 1859 raid on a U.S. armory just before the outbreak of the Civil War?
    Harper's Ferry

35. Amundsen, Scott, and Byrd all explored which continent?
    Antarctica

36. Leningrad is to St. Petersburg as Constantinople is to what?
    Istanbul

37. Which country did not have colonies in Africa – Belgium, Portugal, or North Korea?
    North Korea

38. When Mount Vesuvius erupted in A.D. 79, it buried the cities of Herculaneum, Stabiae, and which other Italian city?

Pompeii

39. In July 1998 Russia planned to bury the remains of the last tsar, Nicholas II, and his family in the former imperial capital city. Give the present-day name of this city.
St. Petersburg

40. In 1858, U.S. President James Buchanan and Queen Victoria were among the first to exchange a transatlantic greeting using what means of communication – telephone or telegraph?
Telegraph

41. Tripolitania, Cyrenaica, and Fezzan are historic regions in what present-day African country?
Libya

42. The ruins of Chichén Itzá provide archaeologists with information about the Toltec and which other pre-Columbian civilization?
Maya

43. Name the landlocked country in eastern Africa that was once known as Abyssinia.
Ethiopia

44. Which country in southwest Africa was once a center of the slave trade to Brazil and won its independence from Portugal in 1975 – Angola or Republic of the Congo?
Angola

The Geography Bee Ultimate Preparation Guide

45. Which West African country did NOT take its name from an ancient African kingdom – Senegal, Ghana, or Benin?
    Senegal

46. Which country has a smaller population today than it had in the 1840s – Switzerland or Ireland?
    Ireland

47. Vietnam's history and culture have been influenced by the fact that it was controlled by what European country from 1883 to 1954?
    France

48. The people of which ancient civilization are credited with building the most extensive network of paved roads – Roman or Egyptian?
    Roman

49. In 1741, explorers discovered large populations of sea otters and fur seals in the waters around the Aleutian Islands. The explorers claimed the area for what country?
    Russia

50. When Muhammad, the Muslim Prophet, left Mecca, he traveled north to what Saudi Arabian city that is now sacred?
    Medina

51. Located between Iraq and Saudi Arabia, what Persian Gulf country has been ruled by the al-Sabah dynasty since the 18th century?

Kuwait

52. Casablanca and Tripoli are historic cities on the northern coast of what continent?
Africa

53. The city of Reykjavík originated on the site of the first Viking farmsteads on what island country?
Iceland

54. What present-day island country in the Gulf of Guinea is believed to have been uninhabited when it was discovered by the Portuguese in the 1400s?
Sao Tome and Principe

55. Pirate ships from the Barbary States harassed merchant vessels from Europe until the mid-1800s. The northern coast of what continent was home to these pirates?
Africa

56. At what Virginia settlement were British forces defeated by American and French troops, ending the fighting during the Revolutionary War?
Yorktown

57. From the mid-1600s to the mid-1800s, a city on Kyushu was the only Japanese port open to foreign trade. Name this port city.
Nagasaki

58. During the era of the Vietnam War, which major river did most refugees from the war cross to reach safety in Thailand?
Mekong River

59. Which strategically located Mediterranean island was an important British naval base until 1979?
Malta

60. Name the Southeast Asian country that gained independence from the Netherlands in 1949.
Indonesia

61. The Qattara Depression is located in what country that gained its independence from the United Kingdom on February 28, 1922?
Egypt

62. A system of jungle trails used as supply routes by the North Vietnamese in the 1960s and 1970s took its name from the country's leader. Name this trail system.
Ho Chi Minh Trail

63. Cuzco, a city in the mountains of Peru, was the capital of which pre-Columbian empire?
Inca Empire

The Geography Bee Ultimate Preparation Guide

Political Geography

1. Pakistan and India are currently in dispute over what territory – Kashmir or Punjab?
   Kashmir

2. Eritrea gained independence from Ethiopia in what year – 1992 or 1993?
   1993

3. The United States bought Alaska from what country – Canada or Russia?
   Russia

4. Zionism is a political movement that led to the creation of which country in southwest Asia in 1948 – Israel or Jordan?
   Israel

5. The boundary between the United States and Russia passes between which two islands in the Bering Strait – Kuril Islands or Diomede Islands?
   Diomede Islands

6. The partially recognized republic of Abkhazia is northwest of what country with the capital of Tbilisi – Georgia or Armenia?
Georgia

7. Which country currently hosts a larger refugee population – Spain or Jordan?
Jordan

8. In the early 20th century, an international competition selected a design for the federal capital city that replaced Melbourne. Name this capital city?
Canberra

9. The Arab League, which promotes economic cooperation among its members, is made up of countries on which two continents – Africa and Asia or Africa and Europe?
Africa and Asia

10. Since the 1970s, most of the immigrants in the United States have come from other countries in the Americas and which other continent?
Asia

11. Which of the following cities is the oldest city on the Korean Peninsula and the seat of a communist government – Pyongyang or Seoul?
Pyongyang

12. In 1911 a revolution that began in the present-day city of Wuhan overthrew the imperialist government of what country – China or Mongolia?
    China

13. In 1979 the Ayatollah Khomeini declared which southwest Asian country an Islamic Republic after the ruling shah was forced to flee – Saudi Arabia or Iran?
    Iran

14. What city served as Japan's capital for more than a thousand years and now lends its name to an international agreement aimed at reducing greenhouse gas emissions – Osaka or Kyoto?
    Kyoto

15. How many countries are claiming land on Antarctica?
    7 countries

16. An emir is the head of state in which country – Indonesia or Kuwait?
    Kuwait

17. Which Mediterranean principality has been ruled by the Grimaldi family for more than 700 years – Monaco or San Marino?
    Monaco

18. A coup in 1980 prompted a former Portuguese colony to sever its ties with Guinea-Bissau. Name this present-day island country.

Cape Verde

19. Which city was built during the 1960s to become the capital of Pakistan – Islamabad or Rawalpindi?
Islamabad

20. The number of countries that make up the Balkan Peninsula in Europe almost doubled in 1992 as a result of the breakup of which country in Eastern Europe?
Yugoslavia

21. On May 20, East Timor officially gained independence after claiming sovereignty from Indonesia since 1975. Name the capital of this new country.
Dili

22. Which country administered Papua New Guinea prior to its independence in 1975?
Australia

23. In February 2003, the U.S. announced it would begin to admit 12,000 Bantu refugees who have been displaced from a war-torn country located on the Horn of Africa. Name this country.
Somalia

24. Name the two remaining constituent republics of Yugoslavia (From 2002).
Serbia and Montenegro

25. Name the only country in Southeast Asia not colonized by a European power.
Thailand

26. Maharashtra borders the Arabian Sea and is one of the most urbanized states of which Asian country?
India

27. Name the politically divided island in northwestern Europe that has a Protestant majority in the north and a Catholic majority in the south.
Ireland

28. Last summer (2013), Crown Prince Philippe, the Duke of Brabant, became king of a European country after his father abdicated for health reasons. Name this country.
Belgium

29. Which country that borders Myanmar gained its independence from Pakistan in 1971?
Bangladesh

30. The Yalu River forms part of the political boundary between North Korea and what other country?
China

31. The headquarters of several international organizations, including the African Union, are located in a city that was briefly the capital of Italian East Africa. Name this city.
Addis Ababa

32. The border between Turkey and Armenia remains closed in response to a conflict over the disputed area of Nagorno-Karabakh. This disputed area is claimed by what other country that borders Armenia?
    Azerbaijan

33. In March 2004 seven eastern European countries joined NATO, the North Atlantic Treaty Organization. Of these new NATO members, name the one that borders the Russian oblast of Kaliningrad.
    Lithuania

34. What city is the judicial capital of South Africa?
    Bloemfontein

35. The Green Line was in place from 1975-1990 to separate the Christian and Muslim areas in the only national capital city located on the eastern shore of the Mediterranean Sea. Name this city.
    Beirut

36. What country has more time zones – Russia or India?
    Russia

37. Why does the International Date Line make several zig-zags instead of following a straight line – to avoid crossing land or to avoid active volcanoes?
    To avoid crossing land

38. In 1991, Estonia, Latvia, and Lithuania officially gained independence from what former country?

Soviet Union

39. What country, which lost its border on the Red Sea in 1993, was occupied by Italy prior to and early in World War II?
Ethiopia

40. In 2014, the government of India established a new state out of the northwestern part of Andhra Pradesh. Name this new state.
Telangana

41. What word, adapted from the Greek language, refers to a government run by just a few people?
Oligarchy

42. Which of these countries most recently declared their independence from the United Kingdom – Egypt, India, or Botswana?
Botswana

43. Tucson, the second largest city in Arizona, is located in a region of the U.S. that was acquired from Mexico in 1853 through what agreement?
Gadsden Purchase

44. The Institutional Revolutionary Party, or PRI, had been the dominant political party in which North American country throughout most of the 20th century?
Mexico

45. What was the name of the short-lived island republic that united with Tanganyika to form the country of Tanzania in 1964?
Zanzibar

46. What landlocked country was awarded ownership of the Aozou by the United Nations International Court of Justice after years of territorial dispute with its neighbor, Libya?
Chad

47. A city on Baranof Island served as the principal town of Russian America until Alaska became a U.S. territory. Name this city.
Sitka

The Geography Bee Ultimate Preparation Guide

Animal/Environmental Geography

1. Which arctic animal has traditionally played an important role in the economy and lifestyle of the Sami – reindeer or polar bear?
   Reindeer

2. The yak, a kind of long-haired ox, is used as a pack animal in the mountain regions of which continent – Europe or Asia?
   Asia

3. The Bengal Tiger can be found in Nepal, Bangladesh, Burma, Bhutan, and what other country containing the Eastern Ghats – India or Pakistan?
   India

4. Flamingos feed at the salt marshes of Doñana National Park, which lies at the confluence of migratory paths for European and African birds. This park is located along the Gulf of Cádiz in what country?
   Spain

5. What animal that lives south of the Sahara feeds on termites and is nocturnal – Anteater or Aardvark?
   Aardvark

6. The world has 35 species of what famous carnivorous fish with an unusual equine shape – Seahorse or Piranha?
   Seahorse

7. What big cat can be found today in South Asia, East Asia, and Southeast Asia other than the tiger – clouded leopard or snow leopard?
   Clouded Leopard

8. What animal, found in Southern and Southeastern Africa, is social and can be found in herds – Lions or Zebras?
   Zebras

9. What bird, known for is beautiful color and vibrant plumage, can be found in Central and South America – falcon or macaw?
   Macaw

10. Kinkajous can be found in and live in what biome – tropical rainforest or mountain forest?
    Tropical rainforest

11. The gypsy moth caterpillar, which strips trees of their leaves, was accidentally introduced in the 1800s near Cambridge. This city is located on the Charles River in which New England state?
    Massachusetts

12. The wildebeest is to the savanna as the caribou is to the WHAT?
    Tundra

13. What animal is the only nonhuman primate found in Europe, especially on Gibraltar?
    Macaque

14. North America is home to only one marsupial, which is active at night, or nocturnal. Name this marsupial.
    Opossum

15. The binturong is a rare animal with a cat's face, a bear's body, and a tail that's as long as its body. This animal is native to Indonesia, Malaysia, and what Asian region?
    Southeast Asia

16. More kangaroos than people live in Australia – true or false?
    True

17. Weddell seals, the world's southernmost animal, can be found in what continent?
    Antarctica

18. What condor, the largest raptor in the world, can be found exclusively in the mountains and valleys of the longest mountain chain on land in the world – Andean condor or Himalayan vulture?
    Andean Condor

19. A teju, a South American lizard, accompanied a question about its range, which extends from southern Argentina north to Brazil and includes what coastal country located in between?
Uruguay

20. In March 2003, thousands of endangered sea turtles returned to the Bay of Bengal to nest on beaches in the state of Orissa in which country?
India

21. Melville Island, where herds of musk oxen roam, is located north of Victoria Island in which country?
Canada

22. Africanized, or "killer bees," were originally taken to Brazil and accidentally introduced into the wild. These bees entered the U.S. through which U.S. state bordering the Rio Grande?
Texas

23. The westernmost of the Lesser Sundra Islands once provided habitat for a now extinct tiger subspecies that shared its name with the island. Name this island.
Bali

24. The giant tortoises of Aldabra live on an atoll in the island country that lies northeast of the Comoro Islands. Name this country.
Seychelles

25. Red imported fire ants, native to South America, have a painful sting and can be found in parts of the southern U.S. These ants were first seen in Mobile, a city in which state bordering the Gulf of Mexico?
Alabama

26. In the 1960s tropical fish dealers imported the walking catfish from Asia. This catfish, which is able to move across land, now inhabits Lake Okeechobee and other waterways in which state?
Florida

27. Asian long-horned beetles, which probably arrived in wooden pallets shipped from China, have infested many trees in the most populous Midwestern city. This infestation continues to destroy many trees in which city?
Chicago

28. Llamas are to the people of the Andes as yaks are to the people of the what?
Himalayas

29. Penguins swim up to 3,100 miles in a year. Emperor Penguins live in colonies along the Ross Sea on what continent?
Antarctica

30. The Asian tiger mosquito, which can spread diseases to humans, was brought to the U.S. in used tires. This mosquito is now found in most states east of what river, which is over 2,300 miles long?

Mississippi River

31. The capital city southeast of the Tonle Sap is near the edge of current tiger habitat. Name this city.
    Phnom Penh

The Geography Bee Ultimate Preparation Guide

Current Events

1. In April 2015, a massive earthquake occurred in a South Asian country. This earthquake may have altered the height of a famous mountain in this country. Name this country.
   Nepal

2. In April 2015, President Barack Obama shook hands with Raúl Castro at the Seventh Summit of the Americas in Panama City, Panama. Raúl Castro is the president of what country?
   Cuba

3. In April 2015, the 6,572 ft Calbuco Volcano, experienced two eruptions separated by a few hours. This volcano is located approximately 620 miles south of the capital of what country in South America?
   Chile

4. In April 2015, a deadly tornado destroyed much of the town of Fairdale, which is located in the northern part of what U.S. state?
   Illinois

5. In April 2015, the Large Binocular telescope observatory managed to get a glimpse of a large lava lake on Io, one

of Jupiter's moons for the first time ever. This observatory is located in what state bordering California to the west?
Arizona

These are questions that are very similar to those asked in all levels of the National Geographic Bee. You should study Current Events happening in the two months before the bee.

For example, if your State Geographic Bee is on April 1st, you would study global and U.S. current events in January, February, and March. You should also study current events in December, just to be prepared.

It will be the same with the school and national bees. Study all of the current events happening in the three months before the month of the competition.

These are some helpful links to sites you should use for you study of current events, both globally and nationally.

- www.nationalgeographic.com (the best)
- http://www.timeforkids.com/news-archive/world
- http://www.timeforkids.com/news-archive/nation

The Geography Bee Ultimate Preparation Guide

To prepare for current events, go to some of these sites twice a week and just read through the stories. Gain some new geographical information as you read them, and try to formulate questions like the ones found in the National Geographic Bee to help you prepare. Ask other people to quiz you.

The Geography Bee Ultimate Preparation Guide

National Parks/Forests and Sites

1. Gir Forest National Park, in the state of Gujarat is in what country home to only 411 Asiatic Lions left in the world – India or Nepal?
   India

2. Which Canadian province is home to Banff National Park, a popular tourist stop – Alberta or Manitoba?
   Alberta

3. Which national park is on an island in a lake – Petrified Forest, Isle Royale, or Sleeping Bear Dunes?
   Isle Royale

4. Wayne National Park is located in what U.S. State – Ohio or Michigan?
   Ohio

5. Tongass National Forest is located in what U.S. State – Hawaii or Alaska?
   Alaska

The Geography Bee Ultimate Preparation Guide

6. Angkor is a World Heritage Site in what region of Asia – South Asia or Southeast Asia?
Southeast Asia

7. Scientists study tigers at Way Kumbas National Park on the southeastern coast of Sumatra. What sea borders this park – South China Sea or Java Sea?
Java Sea

8. Known for its cliff dwellings, Canyon de Chelly National Monument is located in which U.S. state – Colorado or Arizona?
Colorado

9. Many plants in Monteverde Cloud Forest Preserve get water both from rainfall and the mist that hangs in the air. This preserve is located south of Nicaragua in what country – Costa Rica or Honduras?
Costa Rica

10. Cradle Mountain-Lake St. Clair National Park, which is located northwest of the city of Hobart, is on an island that is part of what country – New Zealand or Australia?
Australia

11. Royal Chitwan National Park, the site of a tiger-habitat restoration project, is in the foothills of the most populous Himalayan kingdom. Name this country – Nepal or India?
Nepal

12. Lake Wallenpaupack and Allegheny National Forest can be found in what U.S. state – Ohio or Pennsylvania?
Pennsylvania

13. Blackwater National Wildlife Refuge and William Preston Lane Jr. Memorial Bridge are landmarks in what U.S. state bordering the Potomac River – Maryland or Virginia?
Maryland

14. Stellwagen Bank National Marine Sanctuary and Nantucket Sound can be found bordering one U.S. state. Name this state that contains Buzzards Bay – Connecticut or Massachusetts?
Massachusetts

15. Landscape Arch, one of the world's longest natural arches, is found at Arches National Park in which U.S. state – Arizona or Utah?
Utah

16. Amboseli National Park, located north of Mount Kilimanjaro, is in what country that borders Tanzania?
Kenya

17. Mount Cook National Park takes its name from the highest mountain in which island country on the southwestern part of the Ring of Fire?
New Zealand

18. What is the number one visited national park in North America, located in the United States?

Great Smoky Mountains National Park

19. Banff National Park is located in what Canadian province in the western part of the country?
Alberta

20. Which state has more national parks – California, Delaware, or Iowa?
California

21. Where is Chapultepec Park, home to such cultural landmarks as the National Museum of Anthropology, the Museum of Colonial Art, and Maximilian's palace – Tegucigalpa or Mexico City?
Mexico City

22. Gran Paradiso National Park was established to provide protected habitat for ibex, which had been hunted near extinction. This park is south of the Matterhorn in what country?
Italy

23. Santa Cruz, Santa Rosa, and San Miguel are islands within an archipelago that is also a U.S. national park. Name this national park?
Channel Islands National Park

24. Sareks National Park is in one of the most remote parts of which country that borders the Gulf of Bothnia?
Sweden

25. Thousands of mountain climbers and trekkers rely on Sherpas to aid their ascent of Mount Everest. The southern part of Mount Everest is located in which Nepalese national park?
Sagarmatha National Park

26. Geothermal activity caused by diverging tectonic plates creates streaming hot springs in Thingvellir National Park. This national park is in what island country?
Iceland

27. The Torres del Paine National Park is located along the fjord-marked Pacific coastline of which country in the Western Hemisphere – Peru or Chile?
Chile

28. Angkor Wat and Cappadocia are World Heritage sites found on what continent?
Asia

29. Pacific Rim National Park is located on Vancouver Island and is part of which Canadian province?
British Columbia

30. A Russian island that straddles 180 degrees longitude is one of the most biodiverse in the Arctic and is the world's northernmost UNESCO World Heritage Site. Name this island.
Wrangel Island

31. Braslau Lakes National Park is located in the northern region of what country?
Belarus

32. Lake Vanern, Lake Vaitern, and Lake Malaren are all major lakes in what country containing Foteviken Viking Reserve and the Stockholm Archipelago?
Norway

33. Oulanka National Park is close to the Arctic Circle, bordering a Russian National Park. What country is this park located in?
Finland

34. The Tivoli Gardens and Rosenborg Castle are major landmarks in what Scandinavian capital city?
Copenhagen

35. Slowinski National Park can be found in what country that is home to Bialoweiza Forest and borders the Gulf of Gdansk?
Poland

36. A national park in the state of Assam provides habitats for both tigers and one-horned rhinoceroses. What is the name of the river on which this park is located?
Brahmaputra

37. "Equality before the Law" is the motto of a Great Plains state that is home to Chimney Rock National Historic Site. Name this state.

The Geography Bee Ultimate Preparation Guide

Nebraska

38. In which Rocky Mountain state can you find Great Sand Dunes National Park, created in 2004 to protect America's tallest sand dunes?
Colorado

The Geography Bee Ultimate Preparation Guide

Simulation Bee

Classroom/School Competition:

1. Zimbabwe and Botswana are countries found in the southern part of what continent?
   Africa

2. Mumbai is a city in Maharashtra in what country?
   India

3. The Kremlin is located in Moscow, the capital of what country?
   Russia

4. Ontario and Quebec are provinces of what North American country?
   Canada

5. Yemen and Oman are on what peninsula?
   Arabian Peninsula

6. What is the largest country in Africa, located in the Sahara?
   Algeria

7. The Mediterranean Sea borders Europe, Africa, and what other continent?
   Asia

8. Sweden is located on what major peninsula in Europe?

Scandinavian Peninsula

9. North Korea and South Korea are countries found in what Asian geographical region?
East Asia

10. What is the largest lake in Central America?
Lake Nicaragua

11. Tegucigalpa is the capital of what country in Central America?
Honduras

12. Spain borders France to its north and what country to its west?
Portugal

The Geography Bee Ultimate Preparation Guide

State Qualification Test

(Ask your parents or siblings to quiz you)

1. What country produces the most films annually?
a) United States
b) India
c) United Kingdom
d) Brazil

2. Ukraine borders what country to the north?
1) Poland
2) Russia
3) Moldova
4) Belarus

3. The Suez Canal separates Africa from what other continent?
a) Asia
b) Europe
c) North America
d) Australia

4. The Strait of Gibraltar connects the Atlantic Ocean to a smaller sea in the Mediterranean. Name this sea.
a) Bay of Biscay
b) Alboran Sea
c) Black Sea
d) Caspian Sea

5. Madagascar is separated from mainland Africa by what body of water?
a) Chukchi Sea
b) Red Sea
c) Mozambique Channel
d) Arafura Sea

6. Taiwan is separated from Philippines by what strait?
a) Luzon Strait
b) Bering Strait
c) Strait of Dover
d) Strait of Gibraltar

7. The Great Rift Valley can be found on the eastern part of what continent?
a) Asia
b) Europe
c) Africa
d) North America

8. The Yucatan and Baja California Peninsulas can be found in what Central American Country?
a) El Salvador
b) Guatemala
c) Mexico
d) Belize

9. Lake Managua is the second largest lake in Central America, located in what country?
a) Nicaragua
b) Panama

c) Guatemala
d) Belize

10. San Luis Potosi is a city in what country bordering the Gulf of Mexico?
a) Honduras
b) Mexico
c) Nicaragua
d) Costa Rica

11. Nunavut is the largest province of what country?
a) United States
b) Canada
c) Germany
d) Finland

12. St. Pierre and Miquelon is a territory of what country separated from the United Kingdom by the English Channel?
a) Netherlands
b) Luxembourg
c) France
d) Germany

13. Kolkata and Mumbai are major cities in what country containing the Deccan Plateau?
a) India
b) Nepal
c) Thailand
d) Burma

14. Antwerp and Brussels are cities in what country bordering Netherlands?
a) Luxembourg
b) France
c) Germany
d) Belgium

15. Poland borders what sea to the north?
a) Baltic Sea
b) North Sea
c) Norwegian Sea
d) Barents Sea

16. The Wakhan Corridor is located in what South Asian country bordering Iran and Pakistan?
a) Turkmenistan
b) Tajikistan
c) Kyrgyzstan
d) Afghanistan

17. Mandalay is a major city in what South Asian country bordering the Bay of Bengal and Andaman Sea?
a) Cambodia
b) Laos
c) Burma
d) Bhutan

18. Many Hindus in South America can be found in Guyana and what other country with the capital of Paramaribo and borders the Caribbean Sea?
a) French Guiana

b) Venezuela
c) **Suriname**
d) Colombia

19. Durban and Cape Town are cities on a country bordering the Atlantic and Indian Oceans. Name this country.
a) Botswana
b) Namibia
c) Democratic Republic of the Congo
d) **South Africa**

20. Borneo is an island shared by Brunei, Indonesia, and what other country?
a) Vietnam
b) Philippines
c) **Malaysia**
d) East Timor

ESSAY QUESTION

You will write an essay for the final part of the State Qualification Test on your favorite endangered animal. Why is it your favorite? What makes it your favorite? Where is the animal's habitat? What is causing the animal to become endangered?

_____
_____
_____

The Geography Bee Ultimate Preparation Guide

The Geography Bee Ultimate Preparation Guide

State Competition

Round 1: The United States

1. Which state is a bigger producer of cheese – Wisconsin or Idaho?
   Wisconsin

2. Amarillo and El Paso are cities found in what state that borders Mexico and the Gulf of Mexico – Oklahoma or Texas?
   Texas

3. Puget Sound can be found in what western state bordering British Columbia – Washington or Oregon?
   Washington

4. Oahu is an island belonging to what state made up of islands – Hawaii or Alaska?
   Hawaii

5. Chesapeake Bay empties out into what ocean – Atlantic or Pacific?
   Atlantic

6. What state has a shorter coastline on the Gulf of Mexico – Florida or Alabama?
   Alabama

7. What state is made up of two peninsulas – Connecticut or Michigan?
Michigan

8. Rhode Island is south of Massachusetts and east of what state with the cities of New Haven and Waterbury – Connecticut or New York?
Connecticut

9. Oswego, New York is on what Great Lake – Ontario or Erie?
Ontario

10. Minnesota borders North Dakota and what other state to its west – Montana or South Dakota?
South Dakota

11. The Aleutian Islands, in the North Pacific Ocean and the Bering Sea, belong to what state – Hawaii or Alaska?
Alaska

12. What state borders Quebec – Pennsylvania or New Hampshire?
New Hampshire

Round 2: Cultural Geography

The Geography Bee Ultimate Preparation Guide

1. What country in South Asia contains the largest proportion of Hindus in the world?
Nepal

2. Maori is a language spoken in what island country?
New Zealand

3. The Pyramids of Giza, a sacred group of monuments, can be found near what river in Egypt?
Nile River

4. The Zulus, a Bantu ethnic group, are the largest ethnic group in what country?
South Africa

5. Jains can be found in India, the United States, the United Kingdom, and Canada. They are also found in what northeastern African country bordering the Indian Ocean?
Kenya

6. The Kven language, a dialect of Finnish, is spoken in the northern region of what country on the Scandinavian peninsula and that borders Skagerrak?
Norway

7. Urdu and what other language are the official languages of Bihar?
Hindi

8. Tamil, a Dravidian language, is spoken mainly in two countries. Name these countries.
India and Sri Lanka

9. Sundanese is spoken by 40 million people and consists of 15% of the population of what archipelagic country?
   Indonesia

10. A group of nomadic people in Northern Africa wear blue robes. Name this group.
    Tuareg

11. Pueblo de Taos reflects the culture of the Pueblo Native Americans. This settlement can be found in what U.S. state?
    New Mexico

12. What is the name of the religion that spread from India to parts of East and Southeast Asia that has about 550 million followers worldwide?
    Buddhism

Round 3: Physical Geography

1. What is the term that describes a body of water that connects two larger bodies of water?
   Strait

2. What is the term for land that juts out into a body of water and is surrounded by water on three sides?
   Peninsula

3. What is the name of a piece of land surrounded on all sides by water?
Island

4. What term describes an area of highland consisting of flat terrain that is raised above the surrounding area?
Plateau

5. What is the name for a large, slow-moving mass of ice?
Glacier

6. What is the term for a body of water, usually larger than a bay, that is a portion of the ocean cutting into the land?
Gulf

7. What is the term for the distance north and south of the equator?
Latitude

8. What is the term for the distance east and west of the equator?
Longitude

9. The average weather conditions in a certain place over many years is known as what?
Climate

10. What is the name for a chain of islands?
Archipelago

11. A deep, narrow valley that has steep sides is known by what term?
Canyon

12. What is the term for a very steep rock face, usually on the coast and on the side of a mountain?
Cliff

Round 4: Economic Geography

1. What country in South Asia ranks first in world banana production?
India

2. What country, one of the ten largest in the world, leads in bauxite production, ahead of China, Brazil, Indonesia, and India?
Australia

3. What country, containing the city of Durban, is the world's largest producer of platinum?
South Africa

4. What country in East Asia, containing the provinces of Tibet and Inner Mongolia, is the world's largest exporter of computers?
China

5. What country, bordering Pakistan, is the world's second largest producer of onions?
India

6. What country, with the longest coastline in the world, ranks third in aluminum production?
Canada

7. What country that contains Mesa Verde and the city of El Paso ranks third in gold production?
United States

8. What South American country with the cities of Antofagasta and Valparaiso is the world's largest producer of copper?
Chile

9. What country bordering the United States and two oceans ranks second in the world in fluorite production, behind China?
Mexico

10. China and the United States rank first and third in steel production. What country, off the coast of mainland Asia and the birthplace of Shintoism ranks in second?
Japan

11. What landlocked Asian country bordering the Caspian Sea is the world's largest producer of uranium?
Kazakhstan

12. What country in Europe, bordering the Netherlands is the world's largest exporter of cars?
Germany

The Geography Bee Ultimate Preparation Guide

Round 5: The World

1. Tonle Sap is a lake in what country containing the World Heritage Site of Angkor Wat?
Cambodia

2. Kakum National Park is located in what country containing Lake Volta?
Ghana

3. The Fortress of Suomenlinna covers three islands. This fort is located in what Nordic country?
Finland

4. The Pakaraima Mountains and Iwokrama Mountains are in what country containing the Mazaruni river?
Guyana

5. Mount Arabat can be found near the border with Iran and Armenia in what country?
Turkey

6. Lake Biwa is in what country containing the cities of Beppu and Miyazaki?
Japan

7. Lagos is a city on the Atlantic Ocean, in what country home to the Tagus and Douro Rivers?
Portugal

8. The Volcano Observatory is in Rabaul, on the island of New Britain in what country?
   Papua New Guinea

9. The American Highland is located on what continent containing the South Orkney Islands?
   Antarctica

10. Jabalpur is close to the Narmada River, which empties out into the Gulf of Khambhat in what country?
    India

11. Minneriya National Park is located in what country bordering the Palk Strait and the Gulf of Mannar?
    Sri Lanka

12. Namak Lake is located in what country, home to the Elburz Mountains, Persepolis, and Mount Damavand?
    Iran

Round 6: Political Geography

1. Kashmir is a territory claimed by India and what other country?
   Pakistan

2. The separatist states of Abkhazia and South Ossetia were recognized as independent by what large country to the north?

Russia

3. Taiwan is administered by what country in East Asia, bordering Russia?
China

4. The United States bought Alaska for $7.2 million in 1867 from what country?
Russia

5. Yugoslavia was former country in Europe, until it broke up into how many countries?
Seven

6. Armenia and what other country are in dispute over Nagorno-Karabakh?
Azerbaijan

7. East Timor used to be part of what archipelagic country in Southeast Asia?
Indonesia

8. Suriname claims land in the French overseas territory of French Guiana and what country?
Guyana

9. What city is the constitutional capital of Bolivia, located near the Pilcomayo River?
Sucre

10. Bangladesh gained independence from what South Asian country in 1971?
Pakistan

11. What country is known for being the only one in Southeast Asia that was not conquered or colonized by a European country?
Thailand

12. When Eritrea gained its independence, the country it separated from became landlocked. Name this country.
Ethiopia

Round 7: U.S. National Parks

1. Katmai National Park, which protects about 2,000 bears, can be found in what northwestern U.S. state?
Alaska

2. Everglades National Park, named after the Everglades, is located in what state that is also a peninsula?
Florida

3. You can witness the Blue Ridge Mountains and a valley of the same name as the national park from it. Name this national park, located in Virginia.
Shenandoah National Park

4. Isle Royale National Park is known for its many wolves and moose. This national park is on an island in Lake Superior belonging to what state?
Michigan

5. You can find many stalactites in Carlsbad Caverns National Park. This famous national park is located in what state?
New Mexico

6. What national park can be found in Utah and was the first established in this state?
Zion National Park

7. What park, approximately the size of Rhode Island, is located in California, contains giant sequoia trees, and boasts the largest exposed granite monolith in the world, El Capitan?
Yosemite National Park

8. Local Native Americans call Glacier National Park "Backbone of the World". This national park is located in what large state?
Montana

9. What national park in Arizona, famous for its saguaro cactus, the largest in North America, is also a site for hikers and horseback riders?
Saguaro National Park

10. What national park, located in the Gulf of Mexico, is a seven-mile long archipelago of seven low-lying islands and harbors some of the healthiest coral reefs off North American shores?
Dry Tortugas National Park

11. North Cascades National Park is home to a diverse ecosystem as well as snow-topped mountains, rivers, and valleys. This national park is located in what state?
Washington

12. Glaciers, earthquakes, and ocean storms have shaped Kenai Fjords for hundreds of years. Kenai Fjords National Park is located in what state containing Resurrection Bay?
Alaska

The Geography Bee Ultimate Preparation Guide

National Competition Preliminaries

1. Dufourspitze is a peak shared by Switzerland and what other country?
   Italy

2. Cantabria and Navarra are autonomous communities in what western European country?
   Spain

3. Dioula and French are languages in what country that has 60 native dialects?
   Cote d'Ivoire

4. The island of Mindoro is bordered by the Sibuyan Sea in what country?
   Philippines

5. The Vorotan and Debed Rivers both can be found in what country containing Lake Sevan?
   Armenia

6. Vpadina Kaundy is in the Caspian Depression in what country containing the Esil River and the cities of Qyzylorda and Almaty?
   Kazakhstan

7. Lake Eyasi is on the Great Rift Valley in what country?
   Tanzania

8. Mitumba is a mountain range covering much of the eastern portion of what country, home to Lake Mweru on its border with Rwanda?
Democratic Republic of the Congo

9. Moldoveanu is the highest point in what country?
Romania

10. Trelew is a city in what country, home to part of the Parana River?
Argentina

11. Port Elizabeth is a town on the island of Bequia in what country whose natural hazards include hurricanes and volcanoes?
St. Vincent and the Grenadines

12. The Minahasa Peninsula is bordered by the Gulf of Tomimi in what country?
Indonesia

13. The Great Karoo is a plateau region in what country where you can find the Brak and Vaal rivers?
South Africa

14. The Aras river forms part of what country's border with Azerbaijan and Armenia?
Iran

15. The Gulf of Corcovado forms the central coast of what country that owns Isla Sala y Gomez?
Chile

16. In what country can you find people who are nicknamed kiwis after their native bird?
New Zealand

17. What capital city, located in the Etela-Suomen Laani province, is a chief port on the Gulf of Finland?
Helsinki

18. The Tenere Desert, extending into Algeria, can also be found in what country?
Niger

19. Jiquiliesco Bay and the Gulf of Fonseca both border what country containing the Cerron Grande Reservoir and Lake Coatepeque?
El Salvador

20. The Orontes River is located in what country whose major cities include Aleppo and Latakia?
Syria

21. Angoumois and Bourbonnais are regions in what country containing the Jura Mountains and Gironde Estuary?
France

22. Viti Levu and Vanua Levu are the largest islands in what country bordering the Kadavu Passage and located in Melanesia?
Fiji

23. People mainly speak Sinhala, Tamil, and English in what country whose highest point is Pidurutalagala?
Sri Lanka

24. The Chota Nagpur Plateau can be found at the eastern end of the Satpura Mountain Range in what country?
    India

25. Ludhiana is located in the northern region of what country containing the states of Kerala, Andhra Pradesh, Rajasthan, Bihar, and Tamil Nadu?
    India

The Geography Bee Ultimate Preparation Guide

USA Geography Olympiad/iGeo Resources

This chapter is designed to help you prepare for the USA Geography Olympiad and the International Geography Olympiad.

What you'll need to know:

World Geography

U.S. Geography

Physical Geography

Cultural Geography

Economic Geography

Historical Geography

Current Events

Political Geography

Topography and Elevations

Interpreting Maps

Using Geographic Diagrams

What resources you'll need with you:

- Atlas (National Geographic Atlases are the best)
- World Maps (Again, national geographic – go to http://education.nationalgeographic.com/education/mapping/outline-map/?ar_a=1)
- U.S. Maps (You can find this in link provided and NG Atlases
- State Maps (United States, Canada, Australia, etc – you can find these in link provided and in NG Atlases)
- Blank Maps
- Current Event websites with geo-related information (Try nationalgeographic.com, dogonews.com, and timeforkids.com (world))
- Earth Science textbooks (Borrow them from your library or school)

The Geography Bee Ultimate Preparation Guide

There are more than 1000 questions here to help you prepare. These are not included as part of the questions found in this book.

So you have 1,930 questions from The Geography Bee Ultimate Preparation Guide, and now another thousand! Study hard!

Links:

http://www.geographyolympiad.com/regionals/qualifying-exams-answer-keys/

http://www.geographyolympiad.com/nationals/sample-questions-geography-challenge-nationals/

http://www.geoolympiad.org/fass/geoolympiad/previous.shtml

The Geography Bee Ultimate Preparation Guide

Geo Statistics

Largest Country in the World: Russia

Smallest Country in the World: Vatican City

Most Populous Country in the World: China

Least Populous Country in the World: Vatican City

Largest Continent in the World: Asia

Smallest Continent in the World: Australia/Oceania

Most Populous Continent in the World: Asia

Least Populous Continent in the World: Australia/Oceania

Largest Island by Area: Greenland

Highest Point in the World: Mount Everest

Lowest Point in the World: Mariana Trench (Challenger Deep)

Lowest Surface Point in Asia: Dead Sea

Lowest Surface Point in Africa: Lake Assal

Lowest Surface Point in South America: Laguna del Carbon

Lowest Surface Point in North America: Death Valley

Lowest Surface Point in Europe: Caspian Sea

Lowest Surface Point in Australia/Oceania: Lake Eyre

Lowest Surface Point in Antarctica: Byrd Glacier

Highest Point in Asia: Mount Everest

Highest Point in South America: Cerro Aconcagua

Highest Point in North America: Denali

Highest Point in Africa: Kilimanjaro

Highest Point in Europe: El'brus

Highest Point in Antarctica: Vinson Massif

Highest Point in Australia/Oceania: Mount Kosciuszko

Tallest Mountain Above and Below Sea Level: Mauna Kea

Highest Mountain Above Sea Level: Mount Everest

Longest Mountain Range Above Sea Level: Andes Mountains

Longest Mountain Range Above and Below Sea Level: Mid-Ocean Ridge

Largest Cave Chamber: Sarawak Chamber, Gunung Mulu National Park, Malaysia

Longest Cave System: Mammoth Cave

Lowest Point in the Pacific Ocean: Mariana Trench (Challenger Deep)

Lowest Point in the Atlantic Ocean: Puerto Rico Trench

Lowest Point in the Indian Ocean: Java Trench

Lowest Point in the Arctic Ocean: Molloy Deep

Largest Ocean: Pacific Ocean

Smallest Ocean: Arctic Ocean

Largest Sea in the World: Coral Sea

Longest River in the World: Nile River

Longest River in Africa: Nile River

Longest River in South America: Amazon River

Longest River in Asia: Yangtze River

Longest River in North America: Mississippi-Missouri River

Longest River in Europe: Volga River

Largest River Drainage Basin in the World: Amazon River

Largest River Drainage Basin in South America: Amazon River

Largest River Drainage Basin in Africa: Congo River

Largest River Drainage Basin in North America: Mississippi-Missouri River

Largest River Drainage Basin in Asia: Ob-Irtysh River

Largest Lake in the World: Caspian Sea

World Population: 7,290,000,000 (7.29 billion)

Most Densely Populated Country in the World: Monaco

Least Densely Populated Country in the World: Mongolia

Countries Sharing the Greatest Number of Borders with other Countries: China and Russia

Tallest Building in the World: Burj Khalifa, Dubai, United Arab Emirates

Tallest Pyramid in the World: Great Pyramid of Khufu, Egypt

Longest Wall in the World: Great Wall of China

Longest Road in the World: Pan-American Highway

Longest Railroad in the World: Trans-Siberian Railroad, Russia

Longest Bridge in the World: Danyang-Kunshan Grand Bridge, China

Longest Suspension Bridge in the World: Akashi-Kaikyo Bridge, Japan

Tallest Road Bridge in the World: Millau Viaduct, France

Largest Reservoir by Surface Area: Lake Volta, Volta River, Ghana

Largest Reservoir by Volume: Lake Kariba, Zambia/Zimbabwe

Tallest Dam in the World: Nurek Dam, Vakhsh River, Tajikistan

Largest Hydroelectric Power Station in the World: Three Gorges Dam, China

Longest Submarine Cable in the World: Sea-Me-We 3 (Southeast Asia – Middle East – Western Europe) cable, connects 33 countries on four continents

Hottest Place in the World: Dalol, Danakil Depression, Ethiopia

Coldest Place in the World: Ridge A, Antarctica

Hottest Recorded Air Temperature: Furnace Creek Ranch, Death Valley, California

Coldest Recorded Air Temperature: Antarctica

Wettest Place in the World: Mawsynram, Meghalaya, India

Driest Place in the World: Arica, Atacama Desert, Chile

Largest Hot Desert in the World: Sahara Desert, Africa

Largest Cold (Ice) Desert in the World: Antarctica

Largest Canyon in the World: Grand Canyon, Colorado River, Arizona

Largest Coral Reef Ecosystem in the World: Great Barrier Reef, Australian Pacific Coast

Greatest Tidal Range in the World: Bay of Fundy, Canadian Atlantic Coast

Tallest Waterfall in the World: Angel Falls, Venezuela

Deepest Lake in the World: Lake Baikal, Russia

Oldest Lake in the World: Lake Baikal, Russia

Strongest Recorded Wind Gust: Barrow Island, Australia

List of Countries by Population:
1. China – 1,350,000,000
2. India – 1,250,000,000
3. United States – 320,000,000
4. Indonesia – 250,000,000
5. Brazil – 205,000,000
6. Pakistan – 200,000,000

7. Nigeria – 180,000,000
8. Bangladesh – 170,000,000
9. Russia – 142,000,000
10. Japan – 127,000,000
11. Mexico – 121,000,000
12. Philippines – 110,000,000
13. Ethiopia – 105,000,000
14. Vietnam – 95,000,000
15. Egypt – 90,000,000
16. Turkey – 84,000,000
17. Iran – 83,000,000
18. Germany – 81,000,000
19. Democratic Republic of the Congo – 80,000,000
20. Thailand – 67,000,000
21. France – 65,000,000
22. United Kingdom – 64,000,000
23. Italy – 62,000,000
24. Burma – 58,000,000
25. South Africa – 55,000,000

List of Countries and Territories by Area:

1. Russia
2. Canada
3. China
4. United States
5. Brazil
6. Australia
7. India
8. Argentina
9. Kazakhstan
10. Algeria
11. Democratic Republic of the Congo

12. Greenland
13. Saudi Arabia
14. Mexico
15. Indonesia
16. Sudan
17. Libya
18. Iran
19. Mongolia
20. Peru
21. Chad
22. Niger
23. Angola
24. Mali
25. South Africa

List of Languages by Native Speakers:

1. Mandarin Chinese – 990,000,000
2. Hindi – 460,000,000
3. Spanish – 405,000,000
4. English – 360,000,000
5. Arabic – 290,000,000
6. Bengali – 240,000,000
7. Portuguese – 215,000,000
8. Russian – 155,000,000
9. Japanese – 125,000,000
10. Punjabi – 100,000,000
11. German – 95,000,000
12. Tamil – 83,000,000
13. Telugu – 82,000,000
14. Javanese – 81,000,000
15. Wu – 80,000,000
16. Malay/Indonesian – 78,000,000

17. Marathi – 77,000,000
18. French – 76,000,000
19. Vietnamese – 75,000,000
20. Korean – 74,000,000
21. Urdu – 70,000,000
22. Turkish – 63,000,000
23. Italian – 60,000,000
24. Cantonese – 57,000,000
25. Gujarati – 55,000,000

List of Languages by Total Speakers:

1. English – 1,500,000,000
2. Mandarin Chinese – 1,100,000,000
3. Hindi – 700,000,000
4. Spanish – 450,000,000
5. French – 370,000,000
6. Arabic – 300,000,000
7. Russian – 275,000,000
8. Bengali – 250,000,000
9. Portuguese – 230,000,000
10. German – 180,000,000
11. Malay/Indonesian – 160,000,000
12. Urdu – 155,000,000
13. Japanese – 127,000,000
14. Persian – 115,000,000
15. Telugu – 110,000,000
16. Tamil – 105,000,000
17. Punjabi – 100,000,000
18. Javanese – 87,000,000
19. Turkish – 83,000,000
20. Marathi – 82,000,000
21. Korean – 78,000,000

22. Vietnamese – 75,000,000
23. Wu – 74,000,000
24. Italian – 70,000,000
25. Gujarati – 65,000,000

List of Religions by Number of Followers:

1. Christianity – 2,200,000,000
2. Islam – 1,600,000,000
3. Hinduism – 1,150,000,000
4. Buddhism – 550,000,000
5. Shenism (Chinese Folk) – 450,000,000
6. African Traditional – 110,000,000
7. Shintoism – 100,000,000
8. Sikhism – 35,000,000
9. Taoism – 30,000,000
10. Judaism – 14,000,000
11. Korean Shamanism – 10,000,000
12. Jainism – 8,000,000
13. Confucianism – 7,000,000
14. Caodaism – 6,500,000
15. Baha'i Faith – 6,000,000
16. Cheondoism – 5,000,000
17. Tenriism – 4,200,000
18. Zoroastrianism – 4,000,000
19. Hoahaoism – 3,000,000
20. Rastafarianism – 50,000

Language Families by Population:

1. Indo European – 2,910,000,000
2. Sino-Tibetan – 1,270,000,000

The Geography Bee Ultimate Preparation Guide

3. Niger-Congo – 420,000,000
4. Afro-Asiatic – 350,000,000
5. Austronesian – 320,000,000
6. Dravidian – 300,000,000
7. Turkic – 200,000,000
8. Austroasiatic – 170,000,000
9. Japonic – 130,000,000
10. Tai-Kadai – 100,000,000
11. Koreanic – 80,000,000
12. Uralic – 27,000,000
13. Quechuan – 10,000,000
14. Hmong-Mien – 9,500,000
15. Mongolic – 7,500,000
16. Mayan Languages – 7,000,000
17. Kartvelian – 5,500,000

Most Populous U.S. States:

1. California – 38,000,000
2. Texas – 27,000,000
3. New York – 21,000,000
4. Florida – 20,000,000
5. Illinois – 13,500,000
6. Pennsylvania – 13,000,000
7. Ohio – 12,000,000
8. Georgia – 10,000,000
9. North Carolina – 9,900,000
10. Michigan – 9,000,000
11. New Jersey – 8,900,000
12. Virginia – 8,500,000
13. Washington – 7,500,000
14. Massachusetts – 6,700,000
15. Arizona – 6,500,000

Least Populous U.S. States:

1. Wyoming
2. Vermont
3. Alaska
4. North Dakota
5. South Dakota
6. Delaware
7. Montana
8. Rhode Island
9. New Hampshire
10. Maine

Largest Canadian Provinces by Area:

1. Nunavut
2. Quebec
3. Northwest Territories
4. Ontario
5. British Columbia

Smallest Canadian Provinces by Area:

1. Prince Edward Island
2. Nova Scotia
3. New Brunswick
4. Newfoundland and Labrador
5. Yukon

Most Populous Canadian Provinces:

1. Ontario – 13,000,000
2. Quebec – 8,000,000
3. British Columbia – 4,500,000
4. Alberta – 3,500,000
5. Manitoba – 1,200,000

Least Populous Canadian Provinces:

1. Nunavut
2. Yukon
3. Northwest Territories
4. Prince Edward Island
5. Newfoundland and Labrador

Largest Indian States by Area:

1. Rajasthan
2. Madhya Pradesh
3. Maharashtra
4. Uttar Pradesh
5. Jammu and Kashmir
6. Gujarat
7. Karnataka
8. Andhra Pradesh
9. Orissa
10. Chhattisgarh

Most Populous Indian States:

1. Uttar Pradesh – 199,000,000
2. Maharashtra – 112,000,000
3. Bihar – 103,000,000

4. West Bengal – 91,500,000
5. Tamil Nadu – 72,500,000
6. Madhya Pradesh – 72,500,000
7. Rajasthan – 68,000,000
8. Karnataka – 63,000,000
9. Gujarat – 60,000,000
10. Andhra Pradesh – 55,000,000

Most Populous Metropolitan Areas in the World:

1. Tokyo, Japan – 37,000,000
2. Jakarta, Indonesia – 30,000,000
3. Delhi, India – 25,500,000
4. Seoul, South Korea – 25,000,000
5. Shanghai, China – 24,500,000
6. Manila, Philippines – 24,000,000
7. Mumbai, India – 23,500,000
8. Karachi, Pakistan – 23,000,000
9. Mexico City, Mexico – 22,500,000
10. Guangzhou, China – 22,000,000

Christianity by Country:

1. United States – 230,000,000
2. Brazil – 175,000,000
3. Mexico – 110,000,000
4. Nigeria – 80,000,000
5. Philippines – 78,000,000
6. Russia – 70,000,000
7. China – 65,000,000
8. Democratic Republic of the Congo – 63,000,000
9. France – 55,000,000

10. Italy – 55,000,000
11. Ethiopia – 52,000,000
12. Germany – 50,000,000
13. Colombia – 45,000,000
14. Ukraine – 42,000,000
15. South Africa – 40,000,000
16. Spain – 38,000,000
17. Poland – 37,000,000
18. Kenya – 35,000,000
19. Argentina – 33,000,000
20. United Kingdom – 32,000,000

Islam by Country:

1. Indonesia – 205,000,000
2. Pakistan – 181,000,000
3. India – 161,000,000
4. Bangladesh – 133,000,000
5. Nigeria – 80,000,000
6. Iran – 73,000,000
7. Egypt – 70,000,000
8. Turkey – 70,000,000
9. Algeria – 36,000,000
10. Morocco – 31,000,000
11. Afghanistan – 30,000,000
12. Sudan – 30,000,000
13. Iraq – 29,000,000
14. Ethiopia – 28,000,000
15. Saudi Arabia – 27,000,000
16. Uzbekistan – 25,000,000
17. Yemen – 24,000,000
18. China – 20,000,000
19. Syria – 20,000,000

20. Malaysia – 17,000,000

Hinduism by Country:

1. India – 1,050,000,000
2. Nepal – 25,000,000
3. Bangladesh – 17,000,000
4. Pakistan – 6,000,000
5. Indonesia – 5,000,000
6. Sri Lanka – 3,000,000
7. Malaysia – 1,700,000
8. United States – 1,600,000
9. United Arab Emirates – 1,500,000
10. South Africa – 950,000
11. Burma – 900,000
12. Mauritius – 690,000
13. United Kingdom – 650,000
14. Canada – 500,000
15. Tanzania – 500,000
16. Kuwait – 350,000
17. Fiji – 330,000
18. Australia – 300,000
19. Singapore – 290,000
20. Trinidad and Tobago – 250,000

Buddhism by Country:

1. China – 244,000,000
2. Thailand – 60,000,000
3. Japan – 45,000,000
4. Burma – 40,000,000
5. Sri Lanka – 14,500,000

The Geography Bee Ultimate Preparation Guide

6. Vietnam – 14,000,000
7. Cambodia – 13,500,000
8. South Korea – 11,000,000
9. India – 10,000,000
10. Malaysia – 5,000,000

Sikhism by Country:

1. India – 25,000,000
2. United Kingdom – 900,000
3. Canada – 700,000
4. United States – 520,000
5. Malaysia – 170,000
6. Bangladesh – 120,000
7. Italy – 75,000
8. Thailand – 73,000
9. Burma – 72,000
10. United Arab Emirates – 60,000

Taoism by Country (Estimated):

1. China – 13,000,000
2. Taiwan – 8,000,000
3. Indonesia – 5,000,000
4. Singapore – 630,000
5. Malaysia – 410,000

Judaism by Country:

1. Israel – 6,300,000
2. United States – 5,900,00
3. France – 490,000

4. Canada – 370,000
5. United Kingdom – 290,000
6. Russia – 200,000
7. Argentina – 185,000
8. Germany – 120,000
9. Australia – 100,000
10. Brazil – 95,000

Jainism by Country:

1. India – 6,000,000
2. United States – 90,000
3. Kenya – 80,000
4. United Kingdom – 30,000
5. Canada – 25,000
6. Tanzania – 20,000
7. Nepal – 15,000
8. Uganda – 10,000
9. Burma – 8,000
10. Malaysia – 7,500

About the Author

Keshav Ramesh is a 12-year old author of 16 books and a geography enthusiast. Keshav participated in the 2015 Connecticut State Geographic Bee.

In addition to geography, he participated in the Scripps National Spelling Bee when he was in fourth and fifth grade. Keshav also placed in the top 5% of all sixth graders in the world taking the AMC 8, a math competition.

Keshav participates in statewide piano competitions and previously played in the 12-and-under 2015 USTA Team Tennis. In addition to these competitions, Keshav likes to read, write, explore, play sports, and gain more knowledge. His favorite subjects in school are Math and Social Studies, where he gets to learn more geography/history. Keshav lives in Connecticut with his parents and brother.

You can follow him on Twitter @keshavramesh1

Visit his websites, www.prepgeobee.blogspot.com and www.geobeeworld.blogspot.com for geography bee tips, information, questions, and how to prepare for the Bee.

Acknowledgements

First of all, I would like to thank my parents for buying me my first world atlas when I was six and getting me interested in geography. I would also like to thank them for taking the time to help me prepare, with everything else they were doing for me. They are truly the best!

Second of all, I would like to thank my brother for distracting me while studying geography because I needed to take a little break from it sometimes, and I would forget to.

Thank you to the seventh grade social studies teachers at my middle school, Mrs. Madara, Mrs. Goodale, and Ms. Shattuck for enrolling the school in the National Geographic Bee and giving me a chance to participate and win, ultimately earning me a trip to the 2015 Connecticut State Geographic Bee. I couldn't have gone to the state level if it weren't for them and my principal, Mrs. Larson.

The National Geographic materials were a great help as well. The NG Kids World Atlas, United States Atlas, Ultimate Globetrotting Atlas, and 2015 Almanac were extremely helpful in preparation for the intense competition.

And last but not least, I want to thank all of my friends. From school, I would like to thank Suhas, Harish, and Jonathan for quizzing me with the questions they came up with. Thank you to all of the participants in my school geography bee, my teachers, Lime Team, and my homeroom for cheering me on.

From the Scripps National Spelling Bee, I would like to appreciate Jacob Williamson and Sumedh Garimella for their encouragement. I also want to thank all of my friends from school, the spelling bee, and the geography bee. From the geography bee, I want to thank Karan Menon, Phillip Meng, Saketh Jonnalagadda, Sojas Wagle, Chris Tong, Alan Samineedi, Lucy Chae, and the rest of the GeoClub and GeoBee City communities on Google+.

The Geography Bee Ultimate Preparation Guide

Links

You can use the resources on the next page for geography bee preparation, but here are some that will be vital to your increasing geographic knowledge.

GEOBEE WORLD

www.geobeeworld.blogspot.com

A site created by Keshav Ramesh, Phillip Meng, Karan Menon (2015 NGB Champion), Prithvi Nathan, Alan Samineedi, and Daniel Purizhansky!

THE GOOGLE+ GEOCLUB (Created by Karan Menon and Phillip Meng)

https://plus.google.com/communities/113525651759722425326

THE GOOGLE+ GEOBEE CITY (Created by Karan Menon and Phillip Meng)

https://plus.google.com/communities/109408412585272550345

GEOGRAPHY BEE PREPARATION (Created by Keshav Ramesh

www.prepgeobee.blogspot.com

The Geography Bee Ultimate Preparation Guide

Bibliography

You should use these resources to help you in your preparation for the National Geographic Bee. Use the websites listed here well!

Boyer, Crispin. National Geographic Kids Ultimate U.S. Road Trip Atlas. Washington, D.C.: National Geographic Society, 2012. Print.

Geographic, National. *National Geographic Atlas of the World, Tenth Edition*. N.p.: National Geographic Society, 2014. Print.

Infoplease. Infoplease, n.d. Web. 2015. <http://www.infoplease.com/>.

Kids, National Geographic. *National Geographic Kids Ultimate Adventure Atlas of Earth*. N.p.: National Geographic Society, 2015. Print.

Kids, National Geographic. *National Geographic Kids Ultimate Globetrotting World Atlas*. N.p.: National Geographic Society, 2014. Print.

National Geographic GeoBee. National Geographic, n.d. Web. 2015. <www.nationalgeographic.com/geobee>.

National Geographic Kids Almanac 2016. N.p.: National Geographic Society, 2015. Print.

National Geographic Kids World Atlas. N.p.: National Geographic Society, 2013. Print.

National Geographic Kids United States Atlas. Washington, D.C.: National Geographic Society, 2012. Print.

Wikipedia. Wikimedia Foundation, n.d. Web. 2015. <https://www.wikipedia.org/>.

Wojtanik, Andrew. *The National Geographic Bee Ultimate Fact Book: Countries A to Z*. Washington, D.C.: National Geographic Society, 2012. Print.

"World Geography." *FactMonster World*. FactMonster, n.d. Web. 2015. <http://www.factmonster.com/world.html>.

World Atlas. World Atlas, n.d. Web. 2015. <http://www.worldatlas.com/>.

Notes
UNITED STATES AND THE AMERICAS

ASIA

The Geography Bee Ultimate Preparation Guide

AFRICA

The Geography Bee Ultimate Preparation Guide

EUROPE

AUSTRALIA/OCEANIA AND ANTARCTICA

The Geography Bee Ultimate Preparation Guide

OTHER

The Geography Bee Ultimate Preparation Guide

Made in the USA
Middletown, DE
18 March 2016